SpringerWienNewYork

Gareth R. Eaton • Sandra S. Eaton
David P. Barr • Ralph T. Weber

Quantitative EPR

SpringerWienNewYork

Dr. Gareth R. Eaton
Dr. Sandra S. Eaton
University of Denver
Department of Chemistry & Biochemistry
Denver, CO, USA

Dr. David P. Barr
Dr. Ralph T. Weber
Bruker BioSpin Corporation
Billerica, MA, USA

Printed in Germany

SpringerWienNewYork is a part of Springer Science+Business Media
springer.at

Typesetting: SPI, Pondicherry, India

Printed on acid-free and chlorine-free bleached paper
SPIN: 12540198

With 110 Figures

Library of Congress Control Number: 2010921806

ISBN 978-3-211-92947-6 e-ISBN 978-3-211-92948-3
DOI 10.1007/978-3-211-92948-3
SpringerWienNewYork

Foreword

There is a growing need in both industrial and academic research to obtain accurate quantitative results from continuous wave (CW) electron paramagnetic resonance (EPR) experiments. This book describes various sample-related, instrument-related and software-related aspects of obtaining quantitative results from EPR experiments. Some specific items to be discussed include: selection of a reference standard, resonator considerations (Q, B_1, B_m), power saturation, sample positioning, and finally, the blending of all the factors together to provide a calculation model for obtaining an accurate spin concentration of a sample.

This book might, at first glance, appear to be a step back from some of the more advanced pulsed methods discussed in recent EPR texts, but actually quantitative "routine CW EPR" is a challenging technique, and requires a thorough understanding of the spectrometer and the spin system. Quantitation of CW EPR can be subdivided into two main categories: (1) intensity and (2) magnetic field/microwave frequency measurement. Intensity is important for spin counting. Both relative intensity quantitation of EPR samples and their absolute spin concentration of samples are often of interest. This information is important for kinetics, mechanism elucidation, and commercial applications where EPR serves as a detection system for free radicals produced in an industrial process. It is also important for the study of magnetic properties. Magnetic field/microwave frequency is important for g and nuclear hyperfine coupling measurements that reflect the electronic structure of the radicals or metal ions.

The book contains sufficient background to make it useful to scientists who are new to the field, and who strive to obtain the best possible results from their spectrometers. Information is presented with sufficient rigor so that the book should also be valuable to experienced users who seek to give their work a solid quantitative basis. More detailed discussions and advanced topics are provided in the appendices. Suggestions for further reading guide the reader to additional aspects of EPR. The objective of the authors is to convey this topic to industrial and academic scientists at different stages of their careers.

Acknowledgments

GRE and SSE appreciate the comments made on a draft of this book by Dr. Mark Tseitlin, Dr. Velavan Kathirvelu, and graduate students Deborah Mitchell, Michael Swanson, and Joshua Biller. Richard W Quine and Dr. George A. Rinard have contributed to multiple drafts of the sections on instrumentation, especially measurement of resonator Q and signal-to-noise calculations. Dr. Jinjie Jiang has contributed several experiments that are included in Appendix A.

The development of quantitative EPR in the Eaton lab has been supported by NIH NIBIB grants EB000557 and EB02807 (GRE and SSE), and the Center for EPR Imaging In Vivo Physiology supported by NIH NIBIB P41 EB002034 (Howard Halpern, PI).

Contents

Chapter 1
Basics of Continuous Wave EPR

This chapter is an introduction to the basic theory, practice, and instrumentation for EPR spectroscopy. It is designed to give someone who is new to EPR sufficient background to understand the following chapters. A reader who is already well acquainted with the basics of EPR theory and instrumentation, may skip this chapter. Topics in several sections of this chapter are presented in greater detail in subsequent chapters. This introduction, and much of the book, focuses primarily on molecules with a single unpaired electron, which is denoted as having a spin state of $S = 1/2$.

1.1 The Zeeman Effect

The energy differences studied in EPR spectroscopy are due predominately to the interaction of an unpaired electron in the sample with a magnetic field, B_0, produced by a magnet in the laboratory. This effect is called the Zeeman effect. Because the electron has a magnetic moment, it acts like a compass or a bar magnet when it is placed in a magnetic field. Since energies are quantized, a single unpaired electron has only two allowed energy states. It has a state of lower energy when the moment of the electron, μ, is aligned with the magnetic field and a higher energy state when μ is aligned against the magnetic field (see Fig. 1.1). The two states are designated by the projection of the electron spin, m_s, on the direction of the magnetic field. Because the electron is a spin 1/2 particle, the parallel state has $m_s = -1/2$ and the antiparallel state has $m_s = +1/2$. (Note that because of the charge of the electron, the energies of the $\pm 1/2$ states are opposite to those for the proton.) The difference between the energies of these two states, caused by the interaction between the electron spin and the magnetic field, is shown in (1.1).

$$\Delta E = g\ \mu_B\, B_0\, \Delta m_s = g\ \mu_B B_0, \tag{1.1}$$

where g is the g-factor, which is discussed below, μ_B is the Bohr magneton, which is the natural unit of the electron's magnetic moment, and the change in spin state is

G.R. Eaton et al., *Quantitative EPR*,
DOI 10.1007/978-3-211-92948-3_1,

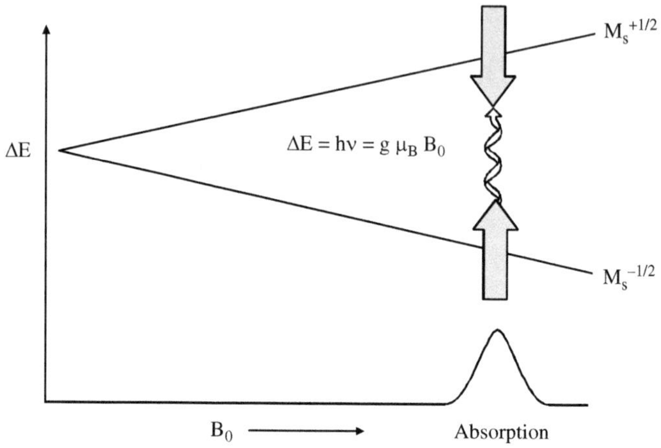

Fig. 1.1 When the magnetic field is scanned, the energies of the two spin states of an unpaired electron diverge. At the magnetic field for which the energy difference between the two electron spin states is equal to hυ for the spectrometer, there is absorption of energy by the spins, which is called resonance

$\Delta m_s = \pm 1$. The energy, $\Delta E = h\nu$, that is required to cause a transition between the two spin states is given by (1.2),

$$\Delta E = h\ \nu = g\ \mu_B B_0 \quad \text{and} \quad \mu_B = g_e\ \beta/2. \tag{1.2}$$

Two facts are apparent from (1.1) and (1.2) and its graph in Fig. 1.1.

- The two spin states have the same energy in the absence of a magnetic field, so without a magnetic field, there is no energy difference to measure.
- The energies of the spin states diverge linearly as the magnetic field increases and the energy difference depends linearly on the magnetic field.

Because the energy difference between the two spin states can be varied by changing the magnetic field strength, there are two potential approaches to obtaining spectra. The magnetic field could be held constant while the frequency of the electromagnetic radiation is scanned, which would be analogous to UV-VIS spectroscopy. Alternatively, the electromagnetic radiation frequency could be held constant while the magnetic field is scanned as shown in Fig. 1.1. Absorption of energy occurs when the magnetic field "tunes" the two spin states such that the energy difference matches the energy of the applied radiation. This field is called the "field for resonance". Because of difficulties in scanning microwave frequencies and because of the use of a resonant cavity for signal detection, most EPR spectrometers operate at constant microwave frequency and scan the magnetic field.

The field for resonance is not a unique "fingerprint" for identification of a compound because spectra can be acquired at different microwave frequencies. The g-factor,

$$g = h\ \nu /\ \mu_B B_0 \tag{1.3}$$

Table 1.1 Field for resonance for a g = 2 sample at various microwave frequencies

Microwave band	Microwave frequency (GHz)	B_0 (for g = 2) gauss
L	1	390
S	3	1070
X	9	3380
K	24	8560
Q	35	12,480
W	94	33,600

is independent of the microwave frequency, so the g-factor is a better way to characterize signals. Note that high values of g occur at low magnetic fields and vice versa. A list of fields for resonance for a g = 2 signal at microwave frequencies commonly available in commercial EPR spectrometers is presented in Table 1.1. Spectrometers are labeled with a letter as shown in the table that is based on the designation that engineers used for microwaves in that frequency range.

The g-factor helps to distinguish and identify types of samples. Carbon-centered radicals have g close to the "free electron value," which is 2.0023. Heteroatoms shift the g-factor. For example benzosemiquinones, which have significant spin density on oxygens, have g ~ 2.004 and nitroxide radicals which have spin density on a nitrogen and oxygen have g ~ 2.006. Metal ions have very different g-factors. For example, the vanadyl ion has g ~ 1.96 while the g-factor for Cu^{2+} may be as large as 2.3, depending on the geometry of the complex. In immobilized samples the anisotropy of g may be observed (i.e. it may depend on the orientation of the molecule with respect to B_0).

1.2 Hyperfine Interactions

Additional information about the species that contains the unpaired electron can be obtained from nuclear hyperfine interactions. The nuclei of the atoms in a molecule or complex often have magnetic moments, which produce a local magnetic field at the electron. The interaction between the electron and the nuclei is called the hyperfine interaction. It gives a wealth of information about the sample such as the identity and number of atoms that make up a molecule or complex; as well as the electron spin density at nuclei that have magnetic moments.

Figure 1.2 depicts the origin of the hyperfine interaction. The magnetic moment of the nucleus acts like a bar magnet (albeit a weaker magnet than the electron) and produces a magnetic field at the electron, B_I. This magnetic field opposes or adds to the magnetic field from the laboratory magnet, depending on the alignment of the moment of the nucleus. When B_I adds to the magnetic field, less magnetic field is needed from the laboratory magnet and therefore the field for resonance is lowered by B_I. The opposite is true when B_I opposes the laboratory field.

For an electron spin interacting with a spin 1/2 nucleus such as hydrogen, the EPR absorption signal splits into two signals that are each B_I away from the original

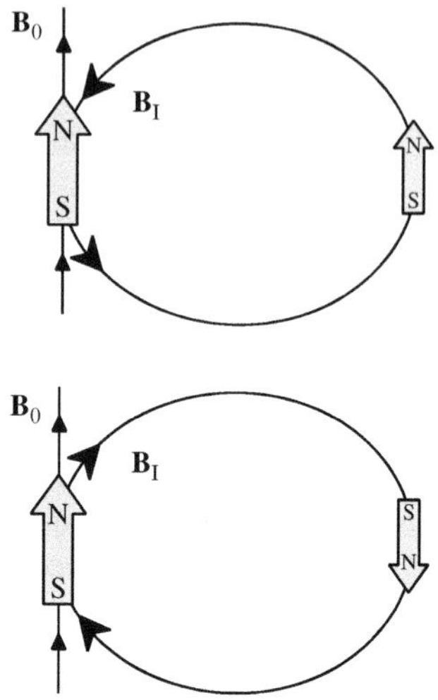

Fig. 1.2 Influence of the magnetic field of a nucleus on the field experienced by an electron spin

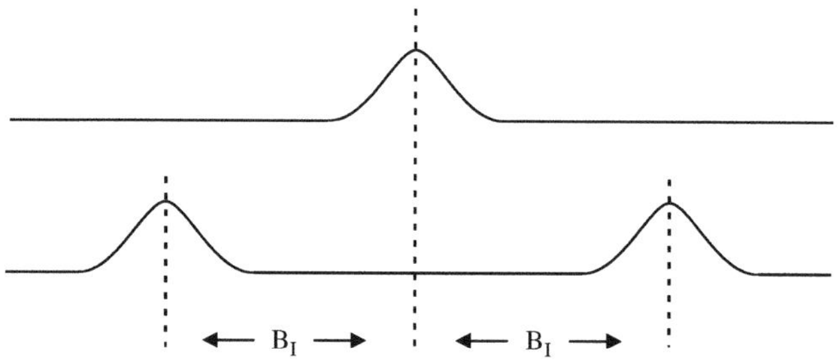

Fig. 1.3 Splitting of an EPR signal due to the local field of a neighboring spin = 1/2 nucleus

signal (see Fig. 1.3). The hyperfine splitting constant, a_H, is equal to the spacing between the two lines, which is $2B_I$.

If there is a second nucleus with I = 1/2, each of the signals is further split into a pair, resulting in four signals. For n spin 1/2 nuclei, there are 2^n EPR signals, provided that the splitting is large relative to the linewidth. The general rule is that the number of hyperfine lines is equal to 2nI + 1, where n is the number of symmetry equivalent nuclei and I is the nuclear spin. Hyperfine splittings can be very useful in identifying radical species. For example, ^{14}N has I = 1, so coupling to one nitrogen gives three lines with equal intensity and Mn has I = 5/2, so coupling to one manganese ion gives six hyperfine lines. As the number of nuclei gets larger, the number of signals increases rapidly. Sometimes there are so many signals that they overlap and a smaller number of broad signals is observed. The magnitude of B_I, and hence the splitting, depends on the type of nucleus, as well as the electron spin density at that nucleus.

1.3 Signal Intensity

So far, only the position of the EPR signal has been discussed. However, the size of the EPR signal is what is important for measurement of the concentration of the EPR active species in the sample. In the language of spectroscopy, the size of a signal is defined as the integrated intensity, i.e., the area beneath the absorption curve (see Fig. 1.4). The integrated intensity of an EPR signal is proportional to the concentration of unpaired electrons in the sample, provided that appropriate operating conditions are used.

Signal intensities do not depend solely on concentrations. They also depend on the microwave power. If low enough microwave power is used, the signal intensity grows as the square root of the power. At higher power levels, the signal grows more slowly, or even diminishes, as well as broadens with increasing microwave power levels. This behavior is called saturation. To measure accurate intensities, linewidths, lineshapes, and closely spaced hyperfine splittings, saturation must be avoided. A quick means of checking for the absence of saturation is to decrease the microwave power and verify that the signal intensity decreases proportional to the square root of the microwave power (for more information on power saturation see Chap. 4.11).

1.4 Introduction to Typical CW EPR Spectrometers

The first half of this chapter provided an introduction to the principles of EPR spectroscopy. Now it is important to consider practical aspects of EPR spectroscopy. Theory and practice have been strongly interdependent in the development and growth of EPR. A good example of this point is the first detection of an EPR signal by Zavoisky in 1945. The Zeeman effect had been known in optical spectroscopy for many years and Gorter had searched for magnetic resonance with a less sensitive method, but the first direct detection of EPR had to wait for improvements in instrumentation. The development of radar during World War II provided technology for rapid development of applications of EPR. Only then, did scientists have the necessary components to build sufficiently sensitive spectrometers. Even today, the availability of improved electronic components has made possible the continued development of advanced techniques in EPR such as Fourier Transform and high frequency EPR. The history of the development of EPR is described in (Eaton et al. 1998).

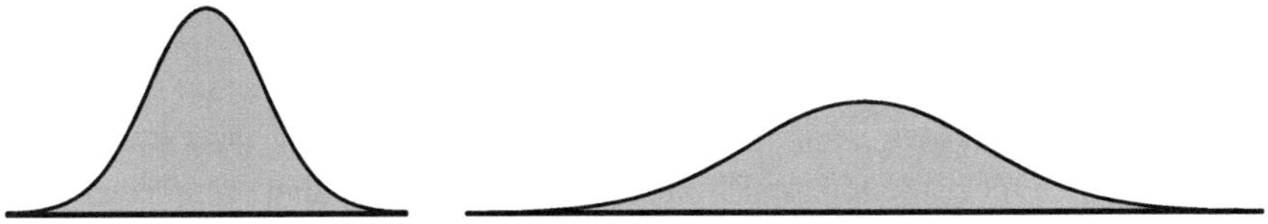

Fig. 1.4 The integrated intensity of the two absorption signals is the same

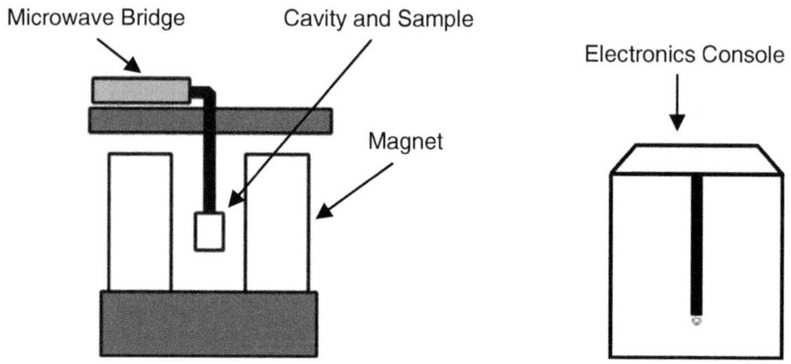

Fig. 1.5 General outlay for an EPR spectrometer

Figure 1.5 shows the general layout of a typical CW EPR spectrometer. The electromagnetic radiation source and the detector are in a box called the "microwave bridge." The sample is in a cavity, which is a metal box that helps to amplify weak signals from the sample. As mentioned in Sect. 1.1, there is a magnet to "tune" the electronic energy levels. In addition, there is a console, which contains signal processing and control electronics and a computer. The computer is used for analyzing data as well as coordinating all the units for acquiring a spectrum. The following sections explain how these parts of the spectrometer function and interact.

1.5 The Microwave Bridge

The microwave bridge houses the microwave source and the detector. There are more parts in a bridge than shown in Fig. 1.6, but most of them are control, power supply, and security electronics that are not necessary to understand the basic operation of the bridge. The following discussion traces the path of the microwaves from the source to the detector.

The tour of the microwave bridge starts at point A, the microwave source. In older spectrometers the source was a vacuum tube called a klystron. Newer spectrometers use a solid state source called a Gunn diode. The output power of the microwave source cannot be varied easily. However the discussion of signal intensity stresses the importance of changing the power level to avoid saturation. Therefore, the next component, at point B, after the microwave source is a variable attenuator: a device that controls the flow of microwave radiation. With the attenuator, the microwave power that gets to the sample can be precisely and accurately controlled.

Most EPR spectrometers are reflection spectrometers. That is, they measure the changes (due to spectroscopic transitions) in the amount of radiation reflected back from the microwave cavity containing the sample (point D in the figure). Therefore the detector should only see the microwave radiation coming back from the cavity.

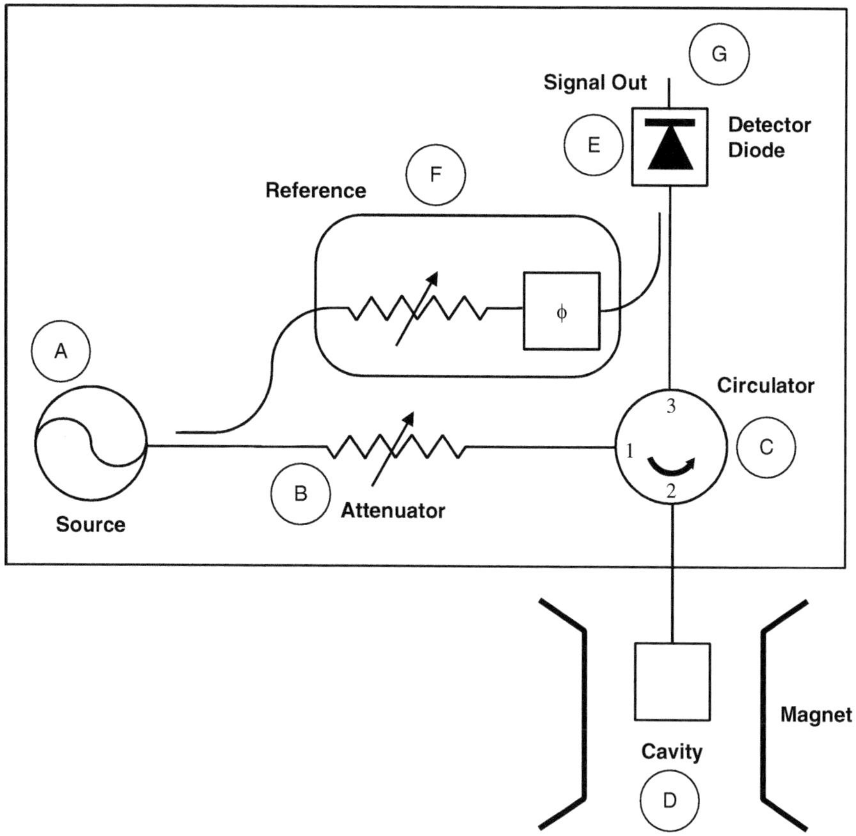

Fig. 1.6 Block diagram for a microwave bridge

The circulator at point C is a microwave device that makes this possible. Microwaves coming in port 1 of the circulator only go to the cavity through port 2 and not to the detector, which is attached to port 3. Reflected microwaves are directed only to the detector (port 3) and not back to the microwave source.

Bruker spectrometers use a Schottky barrier diode to detect the reflected microwaves (point E in Fig. 1.6). The diode converts the microwave power to an electrical current. At low power levels (less than 1 μW), the diode current is proportional to the microwave power, and the detector is called a square law detector. (Remember that electrical power is proportional to the square of the voltage or current.) At higher power levels (greater than 1 mW), the diode current is proportional to the square root of the microwave power and the detector is called a linear detector. The transition between the two regions is very gradual.

For quantitative signal intensity measurements as well as optimal sensitivity, the diode should operate in the linear region. The best results are attained with a detector current of approximately 200 μA. To insure that the detector operates at that level, there is a reference arm (point F in Fig. 1.6) that supplies the detector with extra microwave power or "bias." Some of the source power is tapped off into the reference arm, where a second attenuator controls the power level (and consequently the diode current) for optimal performance. There is also a phase shifter to insure that the reference arm microwaves are in phase with the reflected signal microwaves when the two signals combine at the detector diode.

The detector diodes are very sensitive to damage from excessive microwave power and slowly lose their sensitivity. To prevent this from happening, there is protection circuitry in the bridge that monitors the current from the diode. When the current exceeds 400 μA, the bridge automatically protects the diode by lowering the microwave power level. This reduces the risk of damage due to accidents or improper operating procedures. However, it is good lab practice to follow correct procedures as described later in this book and not rely on the protection circuitry.

1.6 The EPR Cavity

This section discusses the properties of microwave (EPR) cavities and how changes in these properties due to absorption result in an EPR signal. Resonators are used to amplify weak signals from the sample. There is no loss of generality if descriptions are based on a microwave cavity, which is simply a metal box with a rectangular or cylindrical shape that resonates with microwaves much as an organ pipe resonates with sound waves. Many other resonant devices have been used in EPR spectrometers (Poole 1967). Resonance means that the cavity stores the microwave energy. Therefore, at the resonance frequency of the cavity, microwaves remain inside the cavity and are not reflected back (see Fig. 1.7).

Cavities are characterized by their Q or quality factor, which indicates how efficiently the cavity stores microwave energy. As Q increases, the sensitivity of the spectrometer increases. The Q factor is defined as:

$$Q = \frac{2\pi(\mathit{energy\ stored})}{\text{energy dissipated per cycle}}, \tag{1.4}$$

where the energy dissipated per cycle is the amount of energy lost during one microwave period. Energy can be lost to the side walls of the cavity because the microwaves generate electrical currents in the side walls of the cavity which in turn generates heat. Q factors can be measured easily because there is another way of expressing Q:

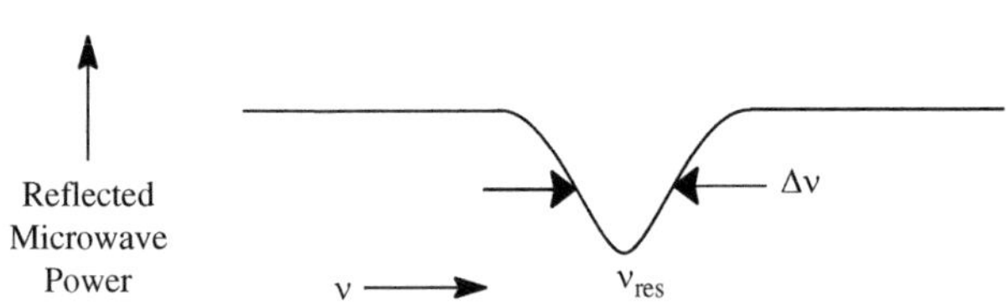

Fig. 1.7 Reflected microwave power from a resonant cavity

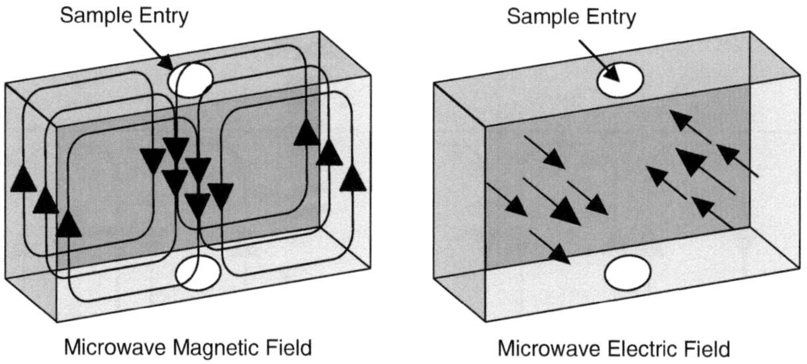

Fig. 1.8 Sketch of the magnetic and electric field patterns in a microwave cavity

$$Q = \frac{\nu_{res}}{\Delta\nu}, \tag{1.5}$$

where ν_{res} is the resonant frequency of the cavity and $\Delta\nu$ is the width at half height of the resonance (see Chap. 7 for more information about resonator Q). A consequence of resonance is that there will be a standing wave inside the cavity. Standing electromagnetic waves have their electric and magnetic field components exactly out of phase, i.e. where the magnetic field is maximum, the electric field is minimum and vice versa. The spatial distribution of the amplitudes of the electric and magnetic fields in the most commonly used EPR cavity (the TE_{102} cavity) is shown in Fig. 1.8. The spatial separation of the electric and magnetic fields in a cavity is used to great advantage. Most samples have non-resonant absorption of the microwaves via the electric field (this is how a microwave oven works) and Q is degraded by an increase in the dissipated energy.

The microwave magnetic field drives the absorption in EPR. Therefore, if the sample is placed in the microwave electric field minimum and the magnetic field maximum, the biggest signals and the highest sensitivity is achieved. Cavities are designed to provide optimal placement of the sample.

Microwaves are coupled into the cavity via a hole called an iris. The size of the iris controls how much of the microwave power is reflected back from the cavity and how much enters the cavity. The iris accomplishes this by carefully matching or transforming the impedances (the resistance to the waves) of the cavity and the waveguide (a rectangular pipe used to carry microwaves). An iris screw in front of the iris lets the user adjust the "matching." When the impedances are matched, the resonator is said to be "critically coupled." This adjustment can be visualized by noting that as the screw moves up and down, it effectively changes the size of the iris (see Fig. 1.9).

How do all of these properties of a cavity give rise to an EPR signal? When the sample absorbs microwave energy, the Q is lowered because of the increased losses, and the coupling changes because the absorbing sample changes the impedance of the cavity. Therefore, the cavity is no longer critically coupled, and microwaves are reflected back to the bridge, resulting in an EPR signal.

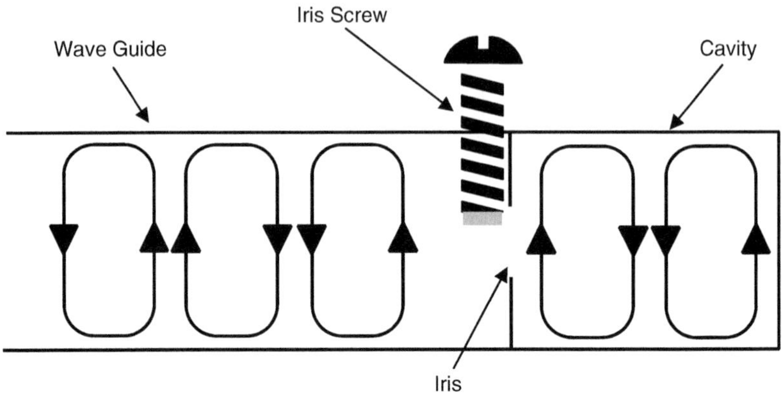

Fig. 1.9 The position of the iris screw relative to the iris is varied to match a microwave cavity to a waveguide

1.7 The Signal Channel

EPR spectroscopists use a technique known as phase sensitive detection to enhance the sensitivity of the spectrometer. This process encodes the EPR signals to make it distinguishable from sources of noise or interference that are almost always present in a laboratory. The encoding helps to decrease the impact of noise from the detection diode and eliminates baseline instabilities due to drift in electronics. The signal channel, a unit in the spectrometer console, contains the electronics required for the phase sensitive detection.

The detection scheme works as follows. The magnetic field strength which the sample sees is varied sinusoidally at the modulation frequency (usually 100 kHz). If there is an EPR signal, the field modulation quickly sweeps through part of the signal and the amplitude of microwaves reflected from the cavity is modulated at the same frequency. For an EPR signal that is approximately linear over an interval as wide as the modulation amplitude, the EPR signal is transformed into a sine wave with amplitude that is proportional to the slope of the signal (see Fig. 1.10).

The signal channel (more commonly known as a lock-in amplifier or phase sensitive detector) produces a DC signal that is proportional to the amplitude of the modulated EPR signal. It compares the modulated signal with a reference signal having the same frequency as the field modulation and it is only sensitive to signals that have the same frequency and phase as the field modulation. Any signals (such as noise and electrical interference) that do not fulfill these requirements are suppressed. To further improve the sensitivity, a time constant is used to filter out more of the noise.

Phase sensitive detection with magnetic field modulation can increase the sensitivity by several orders of magnitude; however, the user must be careful in choosing the appropriate modulation amplitude, frequency, and time constant. A poor choice of any of these three variables can distort the EPR signals and make interpretation of the results difficult.

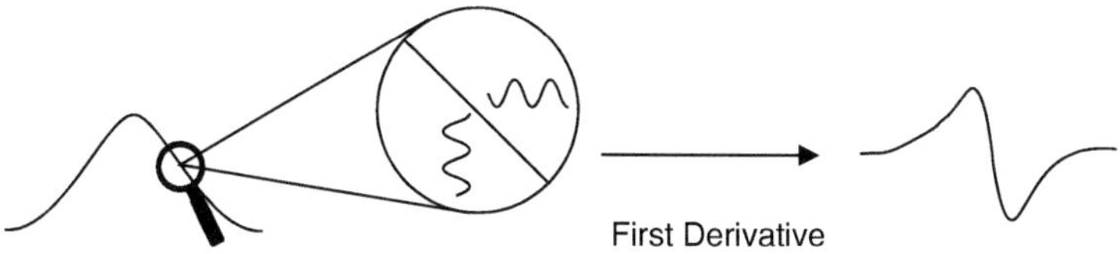

Fig. 1.10 Field modulation and phase sensitive detection

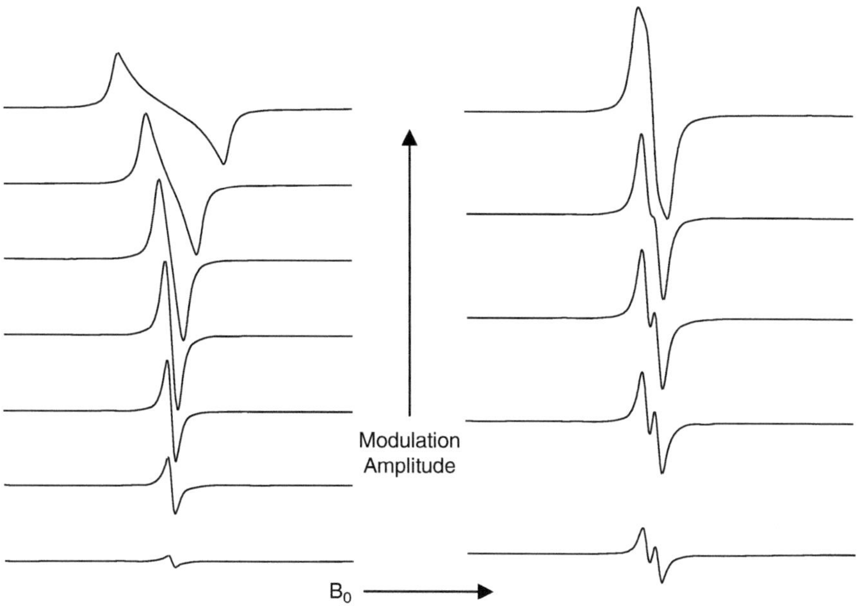

Fig. 1.11 Signal distortions due to excessive field modulation

At low modulation amplitudes, as the amplitude of the magnetic field modulation is increased, the height of the detected EPR signals increases. However, if the modulation amplitude is too large (larger than the line width of the EPR signal), the detected EPR signal broadens and becomes distorted (see Fig. 1.11). The selection of the "best" modulation amplitude depends on the goal of the experiment. To get the most accurate information about signal lineshape, the modulation amplitude should be less than 10% of the distance (in Gauss) between the positive and negative peaks in the derivative spectrum. If this low modulation amplitude results in unacceptably poor signal-to-noise, modulation amplitude can be increased to about 1/3 of the linewidth with very little lineshape distortion. However, although higher modulation amplitudes cause broadening of the signal, the integrated intensity of the signal continues to increase linearly with modulation amplitude so if the goal of the experiment is primarily spin quantitation, higher modulation amplitude may be used to further improve signal-to-noise. A good compromise between signal height and lineshape distortion occurs when the amplitude of the magnetic field

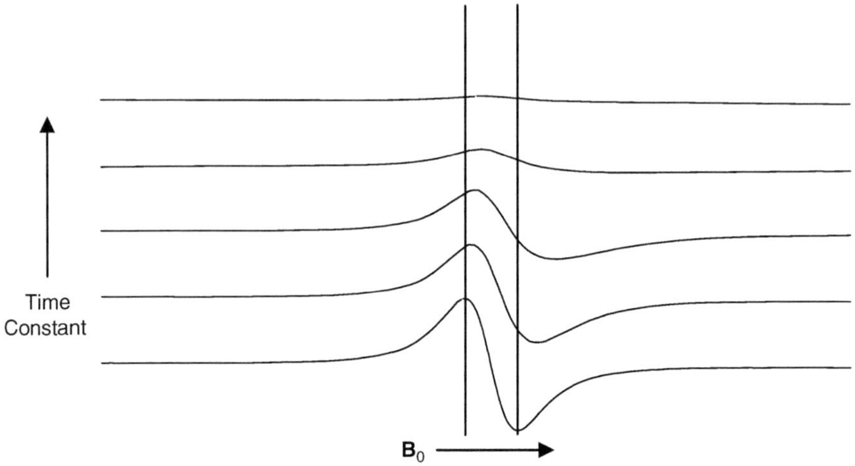

Fig. 1.12 Signal distortion and shift due to excessive time constant

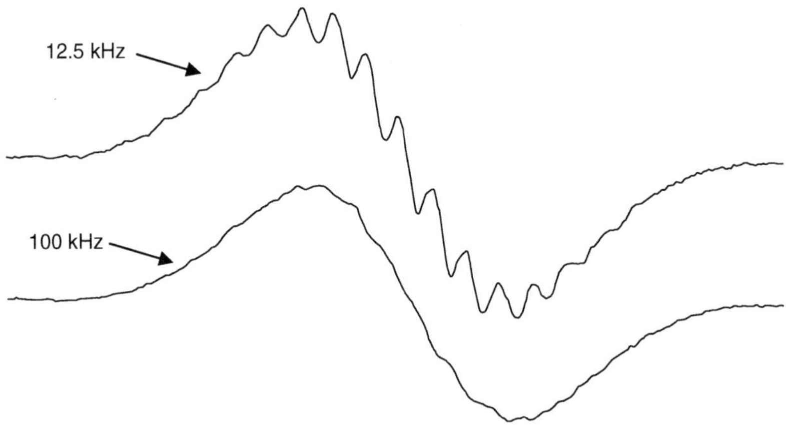

Fig. 1.13 Loss of resolution due to high modulation frequency

modulation is equal to the width of the EPR signal observed at low modulation amplitude. It should also be noted that if the modulation amplitude is greater than the splitting between two EPR signals, the resolution of the two signals is lost. Additional information about modulation amplitude is discussed in Sect. 4.12.

Time constants selected by the operator will affect the noise in the spectrum. Increased time constants filter out noise by slowing down the response time of the detection circuit. As the time constant is increased, the noise level decreases. If a time constant is selected that is too long for the rate at which the magnetic field is scanned, the signal may be distorted or even missed. Also, the apparent field for resonance shifts. Figure 1.12 shows the distortion and disappearance of a signal as the time constant is increased. If a long time constant is needed to distinguish a weak signal from noise, a slower scan rate must be used. A safe rule of thumb is to make sure that the time needed to scan through a single EPR signal should be ten times greater than the time constant.

For samples with very narrow or closely spaced EPR signals (~50 mG), the signal is broadened if the modulation frequency is too high (see Fig. 1.13). The 100 kHz modulation frequency corresponds to 35 mG sidebands on the EPR signal (see Appendix B.4).

1.8 The Magnetic Field Controller

The magnetic field controller defines the center field and sweeps the magnetic field for the scan (Fig. 1.14). It consists of two parts; a digital part that defines the field values and the timing of the field sweep and an analog part that regulates the current in the windings of the magnet to attain the requested magnetic field value. The setting of the magnetic field and the timing of the magnetic field sweep are controlled by a microprocessor in the controller. In older Bruker field controllers a field sweep is divided into a maximum of 4,096 discrete steps (12-bit accuracy) called sweep addresses. Newer instruments will have higher resolution. At each step, a reference voltage corresponding to the magnetic field value is sent to the part of the controller that regulates the magnetic field. The sweep rate is controlled by varying the waiting time between the individual steps.

The magnetic field regulation occurs via a Hall probe placed in the gap of the magnet. It produces a voltage that is dependent on the component of the magnetic

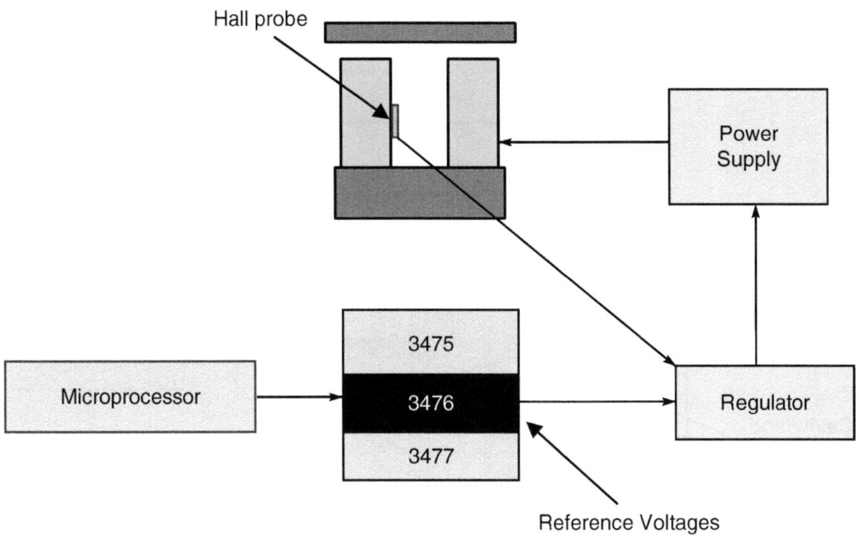

Fig. 1.14 Block diagram for a field controller and associated components

field that is perpendicular to the probe. The relationship between field and voltage is not linear and the voltage changes with temperature; however, this is easily compensated for by keeping the probe at a constant temperature slightly above room temperature and characterizing the nonlinearities so that the microprocessor in the controller can make the appropriate corrections. Regulation is accomplished by comparing the voltage from the Hall probe with the reference voltage given by the other part of the controller. When there is a difference between the two voltages, a correction voltage is sent to the magnet power supply that changes the amount of current flowing through the magnet windings and hence the magnetic field. Eventually the error voltage drops to zero and the field is "stable" or "locked." This voltage comparison occurs at each discrete step of a magnetic field scan.

1.9 The Spectrum

The preceding paragraphs have described the individual components of the spectrometer. Figure 1.15 shows how they work together to produce a spectrum.

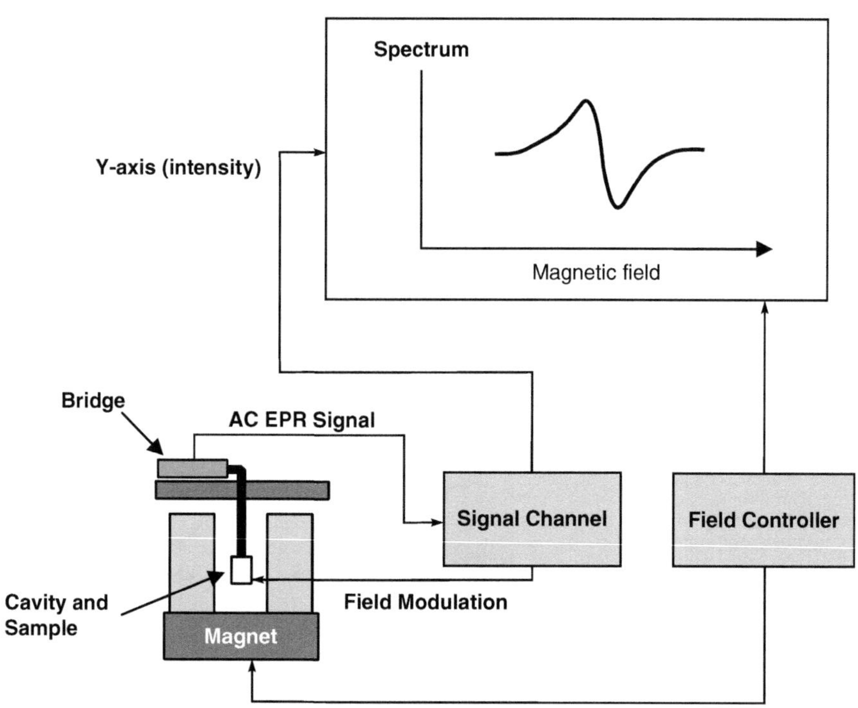

Fig. 1.15 Block diagram for an EPR spectrometer

Chapter 2
Why Should Measurements Be Quantitative?

Even if the question is simply "is there a radical present?" it is important to know, e.g., whether <1 or 100% of the species are in the radical form or in a particular metal oxidation state. There are many examples in the literature in which an impurity or a slight dissociation resulted in the EPR signal observed. A goal of this book is to provide guidance on issues that the EPR spectroscopist should consider when designing experiments to obtain quantitative results. The focus of this book is on radicals in condensed phases.

Among the common type of measurements in which intensity quantitation is essential are:

- How many spins are there in a biological sample?
- What is the spin state of a metal complex as a function of temperature?
- What is the age of an archeological artifact?
- What is the radiation dose?
- What will be the shelf life of foods and beverages?
- How much free radical is produced in certain industrial materials?

Line width quantitation is essential for:

- Oxymetry
- Molecular motion
- Relating line width to relaxation times and hyperfine couplings

Magnetic field quantitation is essential for:

- Measurement of g factors
- Measurement of hyperfine splittings
- Comparison of g or hyperfine with computation of these parameters

G.R. Eaton et al., *Quantitative EPR*,
DOI 10.1007/978-3-211-92948-3_2, © Springer-Verlag/Wien 2010

2.1 Examples of Applications of Quantitative EPR

Burns and Flockhart (1990) reviewed several applications of quantitative EPR, including assays for drugs in body fluids (free radical assay technique, FRAT), radiation dosimetry, molybdenum in sea water, Fe, Mn, and even assays for diamagnetic metals by use of spin-labeled ligands. When the goal is to measure the paramagnetic component of a complicated mixture, EPR may be more selective than other common analytical techniques. An advantage of EPR comes from the fact that it can be applied to samples with scattering properties or opacity that prevent the use of quantitative optical techniques. EPR often does not require any chromatography or other separations techniques before the analysis. For example, samples with a mixture of metals can create a time consuming separations exercise that must be performed before colorimetric, gravimetric or elemental analysis techniques can even be started. EPR can also be applied nondestructively to species such as radiation defects that would not persist through the sample preparation phase for many other types of analytical procedures. Thus, it is not surprising that many reviews of EPR target the analytical community (Saraceno et al. 1961; Molin et al. 1966; Alger 1968; Randolph 1972a, b; Goldberg and Bard 1983; Burns and Flockhart 1990; Eaton and Eaton 1990, 1997; Blakley et al. 2001).

Among the many tools the analyst has available, EPR has some special advantages. For example, one can use several methods to detect the presence of a particular metal, but the EPR g values and hyperfine couplings described briefly in Chap. 1 and in more detail in Chap. 3 provide information about the metal species, not just its presence, thereby permitting speciation as well as quantitation. EPR is very specific to unpaired electrons, and thus is a very good "needle in the haystack" method. It detects only the species with unpaired electrons, ignoring essentially everything else in the sample matrix.

Transition metals with $S = 1/2$ in common oxidation states include V(IV) (especially VO^{2+}), Cr(V), Mo(V), and Cu(II). All of these are easily studied in solids or in solution at room temperature. Although most of the detail in this book is specific to $S = 1/2$ spin systems, most transition metals with odd numbers of unpaired electrons and $S > 1/2$ exhibit readily-observed EPR transitions between $+1/2$ and $-1/2$ spin states. Such common metal ions include $S = 3/2$ Cr(III), $S = 5/2$ Mn(II), $S = 5/2$ Fe(III), and $S = 3/2$ high-spin Co(II). Of the +3 lanthanides, $S = 7/2$ Gd(III) has relaxation times long enough to be readily studied by EPR. The study of even-spin systems, such as $S = 1$ Ni(II), and $S = 2$ high-spin Fe(II), require special EPR techniques and are mentioned only briefly in this book. For more details about even-spin and high-spin species, see Grinberg and Berliner (2004), Möbius and Savitsky (2009), and Weil and Bolton (2007).

One of the most famous early applications of EPR for chemical analysis was the study by Saraceno et al. (1961) of vanadium in petroleum oils. A vanadium porphyrin was used as a standard. Analytical applications of EPR, with a focus on metal ions in solutions, are discussed by Goldberg and Bard (1983). Warren and Fizgerald (1977) used EPR to quantitate Cu(II), Fe(III) and Co(II) ions bound to ion exchange resins. Molybdenum in seawater was determined by converting the Mo to the $Mo(SCN)_5$ complex and extracting into isoamyl alcohol (Hanson et al. (1977). The comparison standard was 2,2,5,5-tetramethyl-3-pyroline-1-oxyl-3-carboxylic acid.

Because of g and hyperfine anisotropies, some species have short relaxation times and broad EPR spectra at room temperature. Sometimes, changing the coordination environment can dramatically sharpen the spectrum. For example, Fe(III) exhibits broad spectra in most coordination environments, but sharpens in the presence of excess F^-. However, in spite of several reports that the EPR of the FeF_6^{3-} complex can be used to analyze for total iron in an aqueous sample (e.g., Burns and Flockhart 1990; Bryson et al. 1975), the equilibrium constant for FeF_6^{3-} formation is not high enough for quantitative analysis. A reproducible analysis could be performed under well-defined and controlled conditions, but a correction for percent complex formation would be necessary for quantitative analysis. This is an example of a case in which the EPR measurement is straight-forward, but the sample preparation is challenging.

Measurement of peroxyl radical concentration permitted the determination of absolute rate constants for the autoxidation of polyolefins (Chien and Wang 1975).

Quantitative EPR has been used to determine concentrations of a very complicated spin system – a mixture of Cr_2O_3 and $NaCrO_2$ (Goldberg et al. 1977). Cr is present as Cr(III) in both species, and both exhibit antiferromagnetic interactions between Cr(III) ions. However, transition to a paramagnetic state occurs at different temperatures in these species, so double integrals of EPR spectra obtained over a range of temperatures revealed relative concentrations after taking into account sample size, temperature, etc.

Concentrations of even diamagnetic ions such as Zn(II) can be determined by EPR if they can be selectively complexed in the sampled matrix with a paramagnetic ligand, or with a ligand that can subsequently be converted to a free radical. The latter case is illustrated by complexing Zn(II) in an aqueous solution with a hydroquinone-containing ligand, which after extraction into an organic solvent subsequently was oxidized to the free radical (Burns and Flockhart 1990).

A very common problem is to determine the extent to which a spin labeling protocol was successful. Did the experiment succeed in adding one and only one nitroxide spin label to a protein or nucleic acid? What is the concentration of spin-labeled lipids in a micelle? Since such answers usually have to be obtained in aqueous samples, obtaining quantitative results implements everything discussed in this book. These problems stimulated studies such as that by Molin et al. (1966).

Whether one is measuring cigarette smoke, protein redox, or beer stability, the answers provided by EPR can be accurate within a few percent, or wrong by more than a factor of two, depending on how carefully one follows good practices as outlined in this book. A few examples of issues that should be considered in designing quantitation experiments are given below.

2.2 Measuring Unstable Radicals by Spin Trapping: Effect of Resonator Q

The steady state concentration of unstable radicals often is too low for direct measurement. Reaction with a spin trap converts these species to more stable trapped radicals. Common spin traps include the pyrroline derivatives DMPO, POBN, and PBN. Blakley et al. (2001) showed that there were large differences in apparent free radical concentration in identical samples depending on the resonator used in the analysis, because the standard and the sample had different effects on resonator Q (see Chap. 7). Adding 0.1 M PBN (*n*-tert-butyl-±-phenylnitrone) to a benzene solution of tempo decreased the resonator Q from 4,400 to 2,600. Only after comparing samples in which radicals were trapped by PBN with standards containing the same concentration of PBN was radical concentration agreement achieved on all three spectrometers tested. Similarly it is also possible to scale the intensity or double integration results by the ratio of the resonator Q-values that is observed from two samples.

2.3 Measuring Weak Signals in the Presence of Strong Ones: Dynamic Range Issues

Distances between unpaired electrons in the range of ca. 4–12 Å can be measured by comparing the relative intensity of the half-field transition to the intensity of the transitions in the g ≈ 2 region (Eaton and Eaton 1982; Eaton et al. 1983). The relevant formula is

$$\text{relative intensity} = \frac{(19.5 \pm 0.5)(9.1)^2}{r^6 \nu^2}, \tag{2.1}$$

where r is the distance in Å and ν is the microwave frequency in GHz.

The relative intensity of the half-field transition is very small. For example, at 8 Å the relative intensity is about 7×10^{-5}. This would be very demanding of the S/N and dynamic range of the spectrometer. However, since the area, and not the line shape, is desired, larger than normal modulation amplitude can be used to improve the S/N. Further, the half-field transition is less easily saturated than is the g ≈ 2 signal, so higher microwave power can be used for the half-field transition. Integration of these signals requires care about background corrections. Paying close attention to these matters, each of which is discussed in more detail later in this book, provides accurate results for an important and useful measurement.

Another aspect of dynamic range in EPR is illustrated by the spectra of radical species derived from C_{60}. After many studies it has been learned that there is a radical impurity in most preparations of C_{60}. In the early days, the spectrum of this radical impurity was incorrectly identified as the spectrum of C_{60}^{-}. The problem

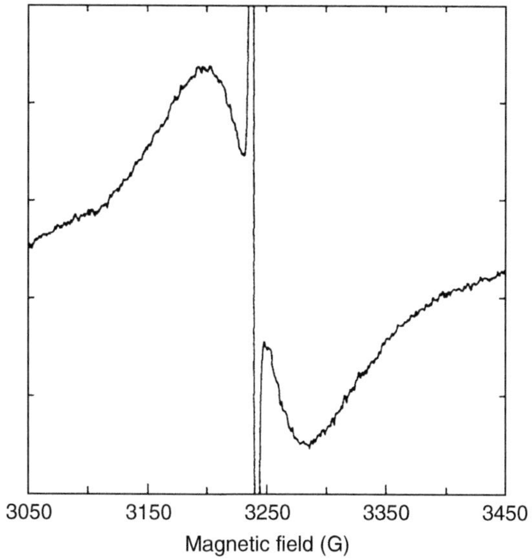

Fig. 2.1 CW EPR spectrum of the C_{60} anion radical with a sharp impurity signal superimposed. From Schell-Sorokin et al. (1992)

was that this spectrum is very narrow, so when it was full scale, the spectrum of the anion was weak enough that it was possible to miss it. Figure 2.1 shows the broad spectrum of C_{60}^{-} when the narrow spectrum is off-scale (Schell-Sorokin et al. 1992).

2.4 Signals in Mixtures

Not all samples to which one might want to apply quantitative EPR are "clean". One interesting problem is to correlate the EPR spectrum in Fig. 2.2 to the age of a blood stain (Fujita et al. 2005). The ratio of the signal labeled H to that labeled g4 provided a linear log–log plot up to 432 days, as shown in Fig. 2.3 (Fujita et al. 2005).

2.5 Radiation Dosimetry

Almost anything used in a hospital operating room is sterilized either with γ rays or e-beam. Many other items, such as food and mail are also treated with ionizing radiation. To ensure sterility (and avoid the legal liability if goods are not sterile) and to run a cost-effective business, the radiation dose measurement must be reliable, reproducible, accurate, and often performed by unskilled laborers. Also the results must be easily transferred to a LIMS (Laboratory Information Management System) for auditing purposes. The preferred method of measuring the radiation dose used for sterilization is EPR radiation dosimetry using alanine dosimeters.

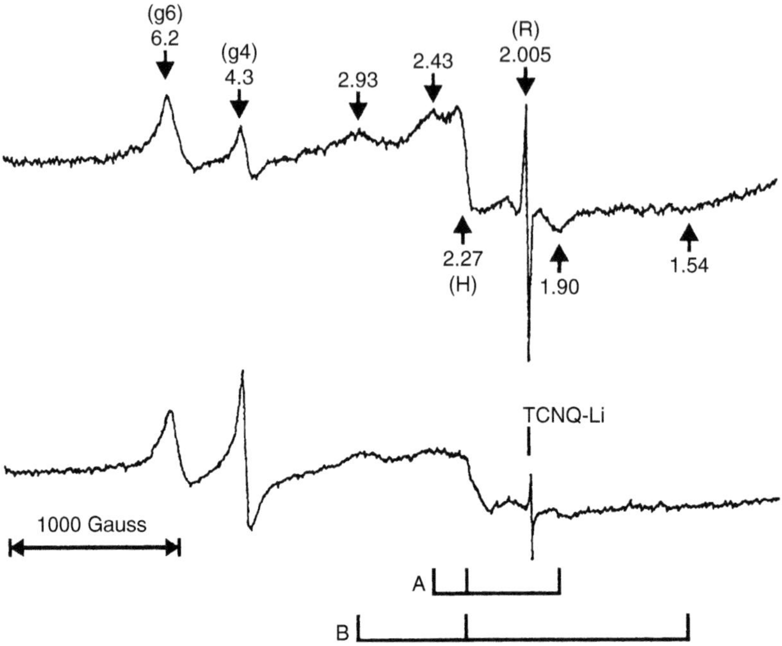

Fig. 2.2 EPR of a blood stain. From Fujita et al. (2005)

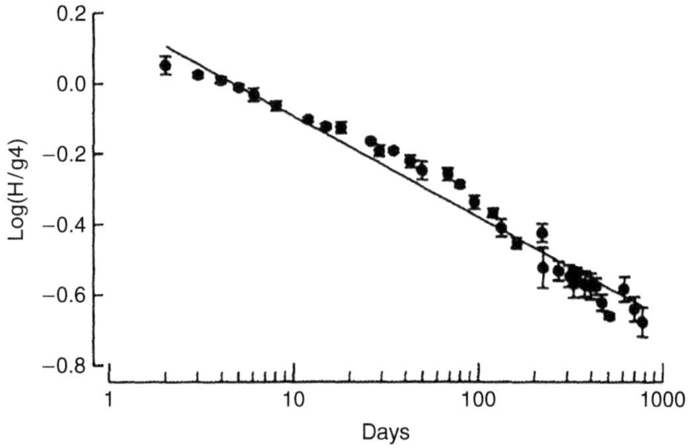

Fig. 2.3 EPR intensity vs. age of dried human blood stains. From Fujita et al. (2005)

The Bruker e-scan benchtop EPR spectrometer has been specifically designed for such applications. The resonator incorporates a special sample holder that reproducibly positions an alanine film or pellet dosimeter in the resonator. The sample holder also has a relative intensity reference standard so that variations in instrument response can be used to normalize the intensity. A bar-code reader is

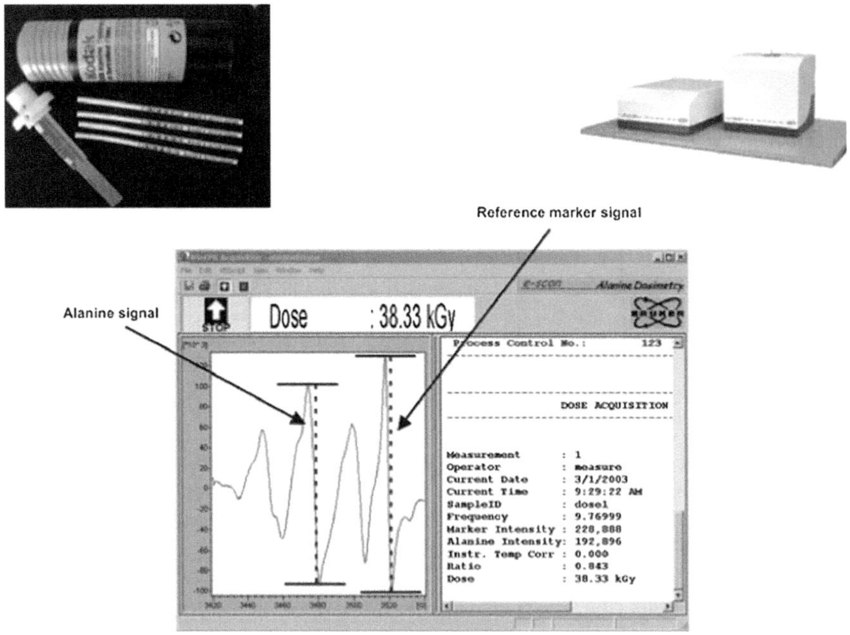

Fig. 2.4 Bruker e-scan with film holder and bar-coded film

incorporated so that the individual dosimeters are identified and the results logged properly into the quality control database.

Through careful control of sample positioning, size, and properties, the instrument provides reproducibly better than 0.5% (Fig. 2.4).

In vivo radiation dosimetry is also an important application of quantitative EPR. For a review of current efforts, see Swartz et al. (2006).

2.6 Use of Accurate Line Width Information

Current effort in many laboratories (refer to the 2004 Workshop on EPR Imaging available on the Bruker BioSpin website) seeks to use quantitative EPR to measure O_2 concentration in vivo. This is a case in which accuracy of line width information is more important than amplitude or g value. Some papers since the 2004 Workshop illustrate successful steps in this field. Halpern and coworkers have carefully calibrated the EPR oximetry method against the standard (more invasive) fluorometric probe method that requires inserting a fiber optic probe into the animal. There is good agreement, and the EPR method has the advantage that it provides three-dimensional information as opposed to just a spot measurement. In addition,

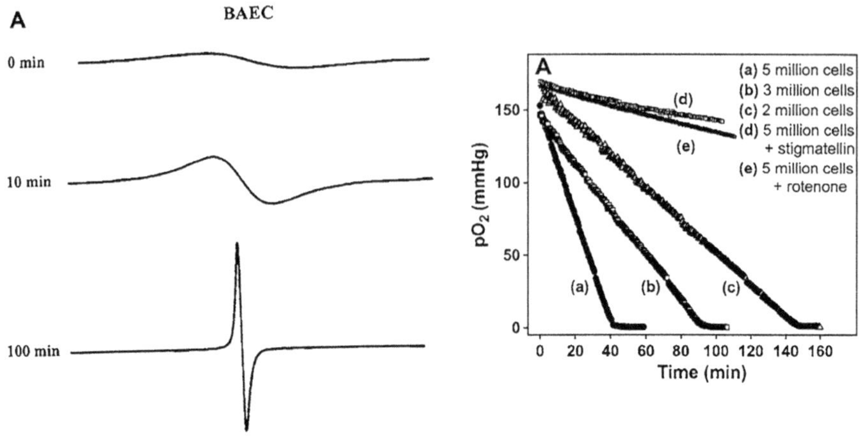

Fig. 2.5 The line width of the X-band EPR spectrum of LiPc microcrystals in a suspension of bovine aortic endothelial cells (BAEC) decreased with time as the cells consumed the oxygen that broadened the EPR signal of the LiPc. The EPR measurements were made with sufficient accuracy to define three phases of cellular respiration. From Presley et al. (2006)

EPR does not damage the tissue that it measures (Elas et al. 2006). Presley et al. (2006) used EPR oximetry to measure cellular respiration (Fig. 2.5).

2.7 Catalysis and Mineralogy

Many applications of quantitative EPR to catalysis and mineralogy were summarized by Dyrek et al. (1994, 2003). Catalytic systems often involve metals in multiple oxidation states and multiple coordination environments. When one of more of these states is paramagnetic, EPR can provide considerable information about the changes in the materials. For example, Sojka and Che (1995) studied species generated during the oxidation of methanol on Mo/SiO_2 catalysts. EPR distinguishes tetra-, penta- and hexa-coordinated Mo(V) species on the SiO_2 substrate, and shows the presence of CH_2OH radical on the surface. When the catalyst was reacted with N_2O, the EPR signal of O^- radicals was observed. The full catalytic cycle was mapped out with the aid of EPR spectra obtained as a function of reagent concentrations and of time at low and room temperature.

2.8 Free Radical Content in Commercial Materials

Free radicals are produced in many industrial processes and can have an adverse effect on the quality of a commercial material. Oxidative reactions caused by light, heat, transition metal catalysis (or any combination of the three) are often unavoidable in a large scale manufacturing method. EPR provides a rapid and simple analytical technique for evaluating the effect that industrial process changes have on the free radical content of the material. The manufacturer can actually use the EPR instrument to optimize their processing methods; then also use EPR as a quality control technique once the optimized method has been put into routine use. In these cases, the operator is not usually concerned with the exact type of free radical that is produced, but merely how much is produced. Thus, relative EPR intensity quantitation of the signal of interest is employed. Relative intensity standards (discussed in Appendix E) help to reduce the uncertainty of the measurements which can often be as low as 1–2%. Some examples of industrial applications that use relative intensity measurements for process and quality control are listed below.

- Shelf life stability in beer
- Rancidity control in vegetable oil based foods
- Irradiation of polymers for use in orthopedic implants
- Oxidation of automotive paint
- Oxidation of pharmaceutical drug excipients
- Free radical production in cigarette smoke

2.9 Feasibility of Quantitative EPR

In this chapter a few of the many applications of quantitative EPR are illustrated. Unfortunately, EPR has previously been judged as inadequately quantitative due to some early inter-laboratory comparisons. Amongst the very practical guidance on EPR techniques in his 1968 book, Alger reported (pp. 202–206) comparative measurements by seven leading laboratories of the number of spins per gram in samples of irradiated sucrose that were circulated among the laboratories and compared with local standards. The range of results between laboratories was as large as 1:1.6. Goldberg (1978) documented that some early EPR spectrometers were not constructed for the degree of precision now desired. In 1994 Yordanov and Ivanova reported comparison by 12 laboratories of samples of Mn(II) and pyrolized sucrose. The results "are characterized by a big variance". These studies should not discourage readers of this book – it is possible, even fairly easy, to obtain good quantitative EPR results. The rest of this book explains how. EPR is based on straight-forward physical principles, and as such, must be quantitatively accurate if the relevant experiment parameters are properly set and controlled. Stimulated by this belief, Dalal et al. (1981a, b) and More et al. (1984) documented the effect of lossy solvents on resonator Q and the effect of solvents, sample tubes, and variable temperature Dewars on B_1 at the sample. The key issue in quantitative EPR is sampling and sample preparation, as is the case for any other analytical method. Beyond that, the interaction of the sample with the spectrometer is important in EPR, so this book provides extensive guidance about that aspect. One should not overemphasize the "uniqueness" of this interaction in EPR. Often, introductions to

EPR and other types of spectroscopy invoke the presumably easily understood concept of transmission visible spectroscopy, so an example from visible spectroscopy is presented here. It is very easy to obtain non-quantitative results in visible absorption spectroscopy with elementary errors of technique, such as non-perpendicular orientation of the sample cell, a small sample that is not in the center of the beam, dirty sample cells, concentration outside the linear range of the spectrometer, and so on. It is the EPR analogs of these elementary mistakes from visible spectroscopy that this book teaches one to avoid.

Unlike early spectrometers, modern commercial EPR spectrometers are designed to provide high levels of linearity and precision in the relevant parameters. Modulation amplitude can be accurately calibrated, and power sensors are incorporated so that the microwave power is known accurately. In the latest spectrometers even the slight variations over the frequency range of the spectrometer are calibrated and stored in the spectrometer. All of the spectrometer settings that this book recommends for quantitative EPR experiments can be included in the software of a spectrometer that is totally computer controlled and calibrated. Thus, the newest spectrometers can include an almost automatic "spin-count". The operator is still responsible for sample preparation and for making sure that all of the assumptions are met and spectrometer parameters are selected appropriately.

To be concise, and precise, the guidance from this book uses specific Bruker hardware/software examples, but the lessons, themselves, are general. Users of other types of EPR spectrometers, or users of earlier or later versions of Bruker hardware and/or software should have little difficulty translating the specifics given in this book to their local hardware/software environment.

2.10 Further Reading

Bolton, Borg, and Swartz (Bolton et al. 1972) summarized selection of experimental parameters, and Randolph discussed quantitative aspects, for the study of biological samples (Randolph 1972a, b).

Chapter 3
Important Principles for Quantitative EPR

The basics of EPR were covered in Chap. 1, for those with no background. Chapter 2 gives examples of quantitative EPR. This chapter discusses in greater depth important principles that one needs to understand when designing a quantitative EPR experiment. Subsequent chapters delve more deeply into topics surveyed in this chapter.

3.1 The EPR Transition and Resulting Signal

As discussed in Chap. 1 for an $S = 1/2$ system, there are two possible spin states $m_s = +1/2$ and $m_s = -1/2$. Transitions between spin states are the basis for EPR spectroscopy (Fig. 3.1). Readers who are more familiar with NMR (or pulsed EPR for that matter), may find that the origin of the signal from a pulsed magnetic resonance experiment is fairly easy to understand. In the usual vector model the B_1 (microwave magnetic field) vector, which is perpendicular to the spin magnetization (taken to be along the z direction), turns the spin magnetization into the xy plane. The magnetization in the xy plane induces a voltage in a detector coil which is detected (i.e., rectified or converted to dc), amplified, and displayed. The magnetization in the xy plane changes with time due to T_2 relaxation (i.e., randomization in the xy plane) and T_1 relaxation (return to equilibrium in the z direction). So, after a single pulse, the resulting induced voltage is due to the free induction decay (FID). After two pulses there is induced voltage due to the FIDs after each of the pulses and additionally from an echo that occurs after the two pulses.

How is this pulsed magnetic resonance vector picture related to the CW experiment? At the field-frequency resonance position, the CW B_1 turns the spins as in the pulse case, but only by a very small amount. A voltage is induced in the conductor of the resonator, as in the pulse case. This voltage is proportional to the angle by which the spins were turned by B_1. Under non saturating conditions T_1 and T_2 are short enough that relaxation back to the z axis is fast relative to the other time constants of the experiment (such as field modulation). Hence, the signal is approximately at

G.R. Eaton et al., *Quantitative EPR*,
DOI 10.1007/978-3-211-92948-3_3, © Springer-Verlag/Wien 2010

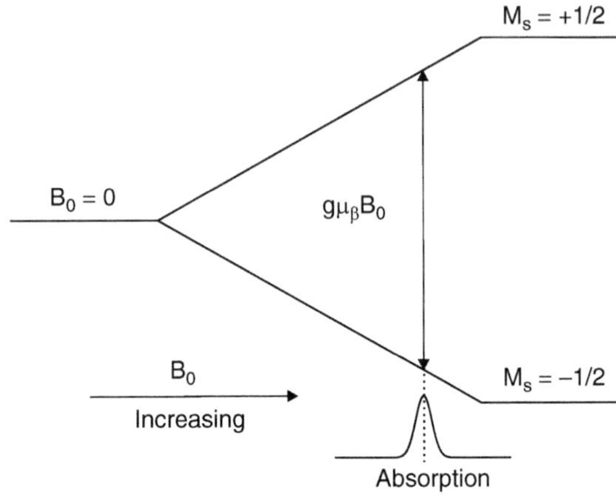

Fig. 3.1 Sketch of the magnetic-field-dependent splitting of the energy levels for a single unpaired electron

equilibrium. However, if the microwave power is too high relative to the relaxation rates, B_1 turns the spins so far from the *z* axis that relaxation cannot return the magnetization back to the *z* axis within the time of the signal measurement. Thus, the induced voltage is less than proportional to B_1, and the detected EPR signal does not increase linearly with the square root of incident power. This is called saturation.

The T_1 relaxation converts microwave energy to heat via the spin–lattice relaxation process. An example would be the electron's magnetic moment coupling with the orbital angular momentum, which in turn, couples to the thermal motion of the molecule. The conversion of microwave energy to heat is akin to microwave energy dissipation in a resistor. As will be shown later in the section on resonator Q (Chap. 7), the resonant absorption of microwaves by the sample has the same effect as a change in resistance of the resonator, and hence results in a lowering of the resonator Q during resonance. By critically coupling the resonator it's impedance is "matched" to that of the microwave transmission line. So, in the absence of resonance there is little or no microwave power reflected back to the detector in the bridge. When microwave power is absorbed during EPR resonance, the resonator becomes mismatched, and power is reflected from the resonator. The reflected power is encoded at the modulation frequency by the magnetic field modulation. In the detection system the reflected power is converted to a voltage that is subsequently amplified, filtered, and displayed. This reflected power is proportional to the magnitude of the EPR absorption by the sample. The magnitude of the EPR absorption is proportional to the amount of spins (or unpaired electrons) in the sample.

3.2 Relaxation and Saturation

As discussed in Chap. 1 for an $S = 1/2$ sample in a magnetic field, the electron spins are characterized by two quantum mechanical states, one with the magnetic moment parallel and the other state with the magnetic moment anti-parallel to the

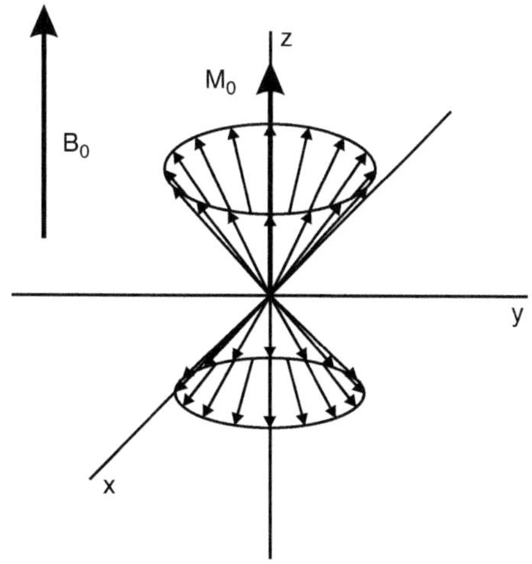

Fig. 3.2 The random orientation of the individual precessing moments cancel their contribution to the transverse magnetization

magnetic field. The moments are almost randomly distributed between parallel and anti-parallel with slightly more in the lower energy parallel state because the electronic system obeys Boltzmann statistics when it is in thermal equilibrium. Then, the ratio of populations of the two states is equal to:

$$\frac{n_{\text{antiparallel}}}{n_{\text{parallel}}} = e^{-\frac{\Delta E}{kT}}, \tag{3.1}$$

where n represents the populations of the two states, ΔE is the energy difference between the two states, k is Boltzmann's constant and T is temperature.

The magnetization that has been discussed so far is actually the vector sum of all the magnetic moments in the sample (Fig. 3.2). Since the moments can only be either parallel or anti-parallel, the magnetization is proportional to the difference, $n_{\text{parallel}} - n_{\text{anti-parallel}}$ and is aligned along the z axis. The difference between the populations of the two spin states is very small. For example at X-band (~ 9.8 GHz) at room temperature (300 K) with a sample that has 10,000 spins, on average 5,004 spins are parallel and 4,996 spins are anti-parallel resulting in a population difference of only 8 (0.08%)! For the same magnetic field strength the population difference for protons is about 600 times smaller, which is why EPR spectra can be obtained at lower spin concentrations than NMR spectra. If the population balance of the sample is perturbed by applying a microwave magnetic field, it is no longer at thermal equilibrium. The absorption of microwave energy heats up the spin system and through the spin system's interactions with the surroundings, it eventually returns to thermal equilibrium. This process is called spin–lattice relaxation and is characterized by the time constant T_1 known as the spin–lattice relaxation time.

M_z is the longitudinal magnetization. M_x and M_y are the transverse magnetization. At equilibrium the longitudinal magnetization is constant, and precesses about

B_0. If there is little interaction between the individual spins of the spin system, the phase of the precession is random and the sum of the individual magnet moments contributing to the transverse magnetization at equilibrium is zero. If there is a net transverse magnetization caused by a perturbation of the system (by applying a microwave magnetic field for example), the system relaxes back to zero transverse magnetization. T_2 is the time constant for this relaxation, which is known as the spin–spin or transverse relaxation.

3.3 Why Are EPR Spectra Displayed as the Derivative?

With a very small net magnetization at X-band, and a small B_1 field continuously perturbing the spin magnetization by only very small amounts, the voltage induced in the resonator in a CW experiment is very small. In order to distinguish this very small signal from noise, magnetic field modulation with phase sensitive detection is employed. Provided that the noise is not correlated with the modulation frequency (usually this is a good assumption), phase sensitive detection greatly enhances the signal-to-noise ratio. The basic principles of phase sensitive are as follows. More details are given in Appendix B.

The magnetic field is modulated sinusoidally as shown in Fig. 3.3 by passing current through a pair of coils mounted on the resonator. The difference $B_{m2} - B_{m1}$ is the modulation amplitude.

If the magnetic field is set to a position within the EPR signal, the oscillating field modulation causes the detected signal to oscillate at the same frequency. The greater the slope of the absorption signal (i.e. a large difference between the absorption signal at the extremes of the modulating field), the greater the amplitude of the oscillating detected signal (Fig. 3.4). This detection method is why CW EPR spectra are the first derivative of the absorption signal.

3.4 Some Caveats About Modulation and First Derivative Displays

If the amplitude of the modulation is larger than the natural line width of the EPR signal, then the spectrum is not the true derivative of the absorption spectrum. In addition, whenever there are two frequencies in a system, there will be sum and difference frequencies. Since the modulation frequency is small relative to the RF/microwave frequency, its signal will occur as "sidebands" on the observed EPR transition. 100 kHz corresponds to ca. 35 mG, so these sidebands are usually well hidden under the envelope of the EPR line. However, if the natural line width of an EPR signal is very narrow, (i.e., $<$35 mG for a 100 kHz modulation frequency) then the sidebands are observed (see Sect. 4.16).

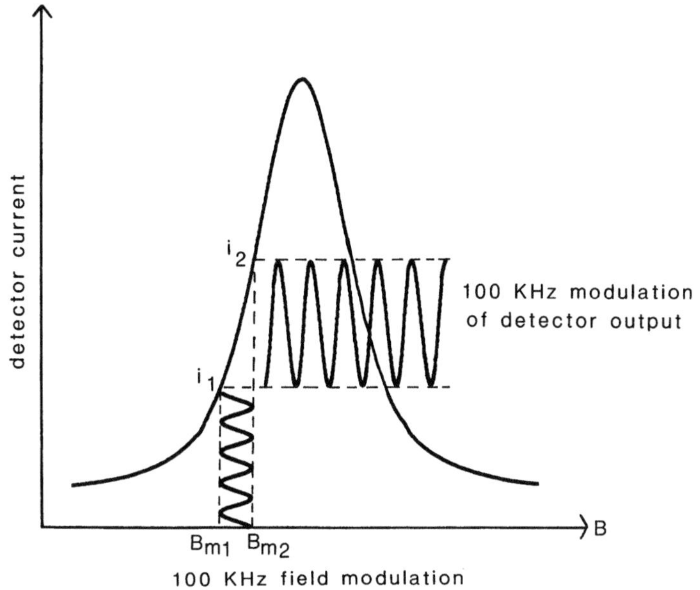

Fig. 3.3 As the main magnetic field is scanned slowly through the EPR line, a small additional oscillating magnetic field, B_m, is applied in the same direction as the main field, B. B_m is commonly at 100 kHz. As B_m increases from the value B_{m1} to B_{m2}, the crystal detector output increases from i_1 to i_2. If the magnitude of B_m is small relative to line width, the detector current oscillating at 100 kHz has a peak-to-peak value that approximates the slope of the absorption curve. Consequently, the output of the 100 kHz phase-sensitive detector is the derivative of the absorption curve. From Eaton and Eaton (2005)

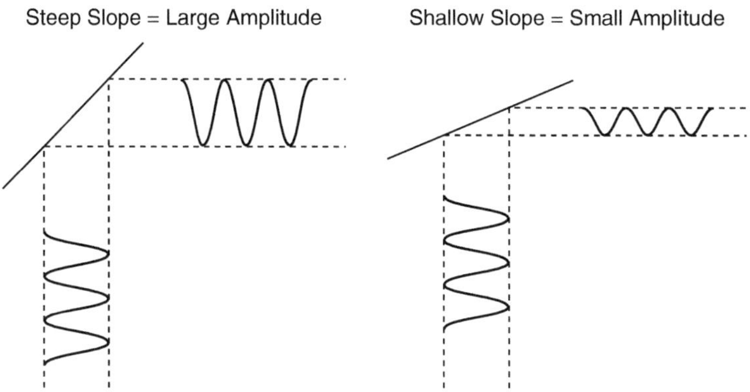

Fig. 3.4 Relationship between the slope of the absorption signal and the detected EPR signal

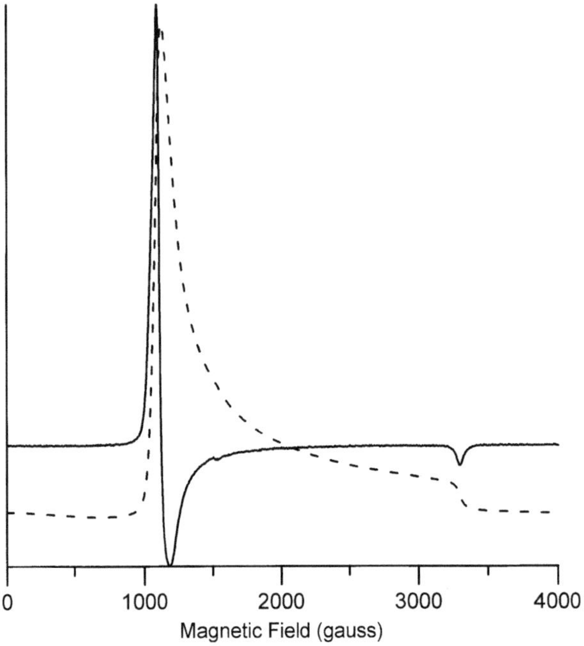

Fig. 3.5 The *dashed line* is an X-band absorption spectrum of a high-spin Fe(III) sample. The *solid line* is the derivative as recorded by the EPR spectrometer. From Eaton and Eaton (2005)

An advantage of derivative spectroscopy is that it emphasizes rapidly-changing features of the spectrum, thus enhancing resolution. However, a slowly changing part of the spectrum has nearly zero slope, so in the derivative display there is "no intensity". An example of how this has entered into the language of EPR is shown in Fig. 3.5 where a high-spin Fe(III) spectrum is displayed. The solid line in this figure is the derivative spectrum as recorded by the EPR spectrometer. It is common to describe this as having "a peak" at ca. $g = 2$ (near 3,300 G in this case) and "a peak" at ca. $g = 6$ (near 1,100 G in this case). However, the absorption spectrum (the dashed line in the figure) has intensity all the way from $g = 6$ to $g = 2$.

3.5 Finding the Signal Area Requires a Double Integration

In many other forms of spectroscopy the initial spectrum is in the form of an absorption curve. In these cases the area under the curve is calculated by integration, and the integral is proportional to the concentration of species present. Since the EPR signal is recorded in the form of a first derivative, double integration is needed to quantify the area under the EPR curve (Fig. 3.6).

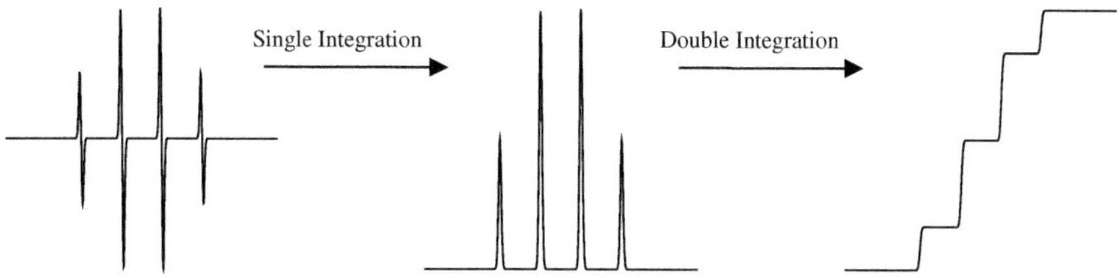

Fig. 3.6 The double integration of an EPR spectrum to reveal the area, which can be used to quantitate the number of spins in the sample

3.6 The CW EPR Line Width

The CW EPR line width of free radicals in fluid solution is usually dominated by unresolved hyperfine interactions. In more viscous solvents and/or at higher field, the line width may be determined by intermediate tumbling rates (Freed 1976; Budil et al. 1996). In frozen solution and in solids, the EPR line width is broadened due to g-factor and hyperfine anisotropy. The distribution of these parameters (often caused by environmental effects) is called g-strain and a-strain.

In other cases the EPR line width is determined by electron spin relaxation; and the line will be Lorentzian in shape. The relation between the derivative peak-to-peak line width, ΔB, and the electron spin relaxation times is given by the following equations:

$$\Delta B^2 = \frac{4}{3}\frac{1}{\gamma^2 T_2^2}\left(1 + \gamma^2 B_1^2 T_1 T_2\right) \tag{3.2}$$

$$\Delta B^2 = \frac{4}{3}\left(\frac{1}{\gamma^2 T_2^2} + B_1^2 \frac{T_1}{T_2}\right) \tag{3.3}$$

$\gamma = 1.7608 \times 10^7$ rad s^{-1} G^{-1}.

Thus, if B_1 is very small, only the first term in parentheses is significant and when T_2 is in seconds this reduces to:

$$\Delta B = \frac{2}{\sqrt{3}\gamma T_2} = \frac{6.56 \times 10^{-8}}{T_2}\ \mathrm{G}. \tag{3.4}$$

This is a very convenient formula for estimating limits on relaxation times. Recall that $T_1 \geq T_2$. For example, when $T_2 = 1$ ... μ, the peak-to-peak derivative line width is 0.0656 G. If the line is relaxation-determined, it is Lorentzian. Note that in this limit T_1 does not affect the line width. Together with the expression for the saturation factor, and an estimate of B_1 in the resonator, this estimate of relaxation times helps select the power to use, as discussed in Sect. 4.11. See Eaton and Eaton (2000b) and Bertini et al. (1994a) for reviews of relaxation times.

However, if B_1 is not small, then the second term in parentheses in (3.3) contributes to the line width, which increases proportional to B_1. Recall that $T_1 \geq T_2$. For samples with longer T_1 values, line broadening will begin to occur at even lower power levels.

Although commercial EPR spectrometers are available at a range of frequencies, most quantitative EPR spectroscopy is likely to be done at X-band. The following sections of this chapter concern issues that arise at higher and lower operating frequencies and/or for samples with $S > ½$. Subsequent sections of the book do not depend on an understanding of these sections.

3.7 Transition Metal EPR

Up to this point in the book only $S = 1/2$ (one unpaired electron) has been discussed. However, whenever there is a possibility for a transition between +1/2 and −1/2 spin states, it is possible to observe an EPR spectrum, so metal ions with an odd number of unpaired electrons are excellent candidates for EPR. This includes many metals in common oxidation states, such as V^{4+}, Cr^{3+}, Mn^{2}, Fe^{3+}, Co^{2+}, Ni^{3+}, and Cu^{2+}, and many clusters, such as iron–sulfur clusters in proteins. If there are closely-spaced excited states, or overlapping higher-spin states, then the relaxation time may be short, in which case cryogenic temperatures may be required to observe the EPR spectrum. It is easy to observe the EPR spectra of Cu^{2+} and VO^{2+} complexes at room temperature, but low-spin Fe^{3+} and Co^{2+} have fast relaxation times at room temperature. For many Co^{2+} complexes, the relaxation time is so long near 4 K that passage effects occur, but the relaxation rate increases proportional to about T^7 and becomes too short to observe the EPR signal above about 10–15 K. For metals with $S > 1/2$ quantitation of EPR signals requires knowledge of which transitions are observed.

3.8 Spectrometer Field and Frequency May Determine Which Transitions Are Observed

For species with $S = 1/2$ at X-band and above, and for common hyperfine splittings, all of the expected lines will be observed and the relative intensities will be as expected. However, at low microwave/RF frequencies and low magnetic fields, some of the expected lines may not be within the accessible magnetic field range, so hyperfine lines will be "missing". Examples are provided by figures in Belford et al. (1987), which are reproduced in Figs. 3.7 and 3.8.

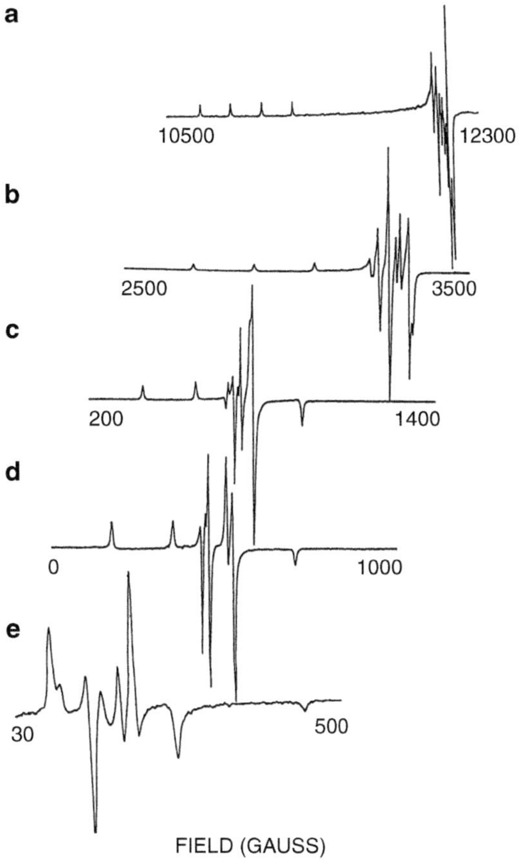

Fig. 3.7 EPR spectra of ^{63}Cu doped into Pd(acac)$_2$ at various microwave frequencies. Magnetic field values are labeled on each scan for the beginning and end of the scan. (**a**) 34.78 GHz, (**b**) 9.376 GHz, (**c**) 2.39 GHz, (**d**) 1.39 GHz, (**e**) 560 MHz. From Belford et al. (1987)

In addition to "missing" lines (see spectrum B in Fig. 3.8), at low frequencies there may appear forbidden lines that become partially "allowed" due to mixing of states.

Recent high-field EPR experiments have shown that many EPR transitions that are not observable at X-band can be detected at higher frequencies. For S > 1/2 systems, if the zero field splitting, D, is >hν (≈ 0.33 cm^{-1} at X-band), some transitions cannot be observed. The field range limit of the spectrometer may also prevent observation of some transitions, even if in principle that could be observed at X-band. For example, Cordischi et al. (1999) pointed out that some Cr(III) transitions were beyond the 8,000 G limit of the spectrometer used.

When guidance is given here or elsewhere (e.g., Eaton and Eaton (1980)) to use a standard as similar as possible to the sample, the intent is to not only compare samples of the same size and solvent, but also standards for which the same type transitions are observed.

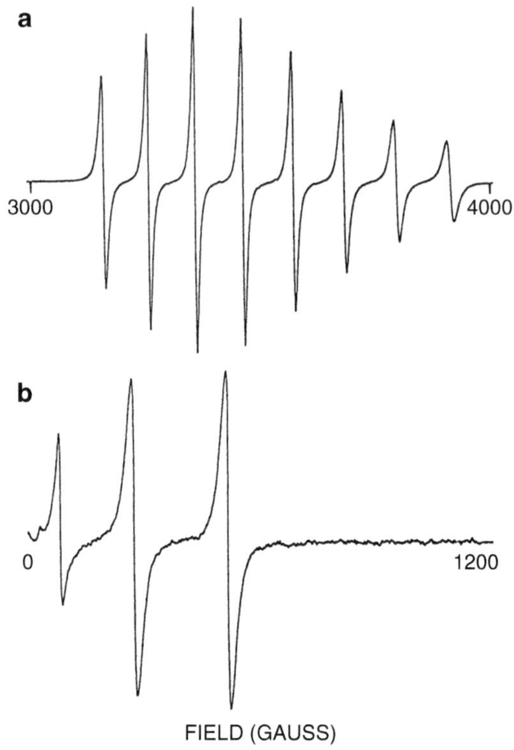

Fig. 3.8 EPR spectrum of $VO(acac)_2$ in 1:1 toluene: chloroform solution at room temperature. Magnetic field values are labeled on each scan for the beginning and end of the scan. (**a**) 9.76 GHz, (**b**) 595 MHz. From Belford et al. (1987)

3.9 Parallel and Perpendicular Transitions

Other common paramagnetic metal ions, such as Fe^{2+} and Ni^{2+}, are missing from the list in Sect. 3.7. In common ligand environments, these ions have even-spin states. Transitions that are forbidden because the spin has to change by more than ± 1 are very weak in the normal EPR spectrometer configurations. Most resonators are carefully designed to have $B_1 \perp B_0$ which permits observation of allowed transitions for half-integer spin systems. However, the transitions in integer spin systems, such as triplets, transitions with nuclear spin flips $\Delta m_I = \pm 1$ (Anderson and Piette 1959), and the "half-field" transitions in interacting spin systems are enhanced with $B_1 \parallel B_0$. The Bruker ER4116DM "dual mode" cavity is designed for these studies. It is called dual mode because it operates in one mode with $B_1 \perp B_0$ at about 9.8 GHz, and in another mode with $B_1 \parallel B_0$ at ca. 9.3 GHz.

Hendrich and coworkers showed that the spectra of integer spin Mn(II) dimers (Figs. 3.9 and 3.10) can be interpreted with one set of parameters when both $\perp$ and $\parallel$

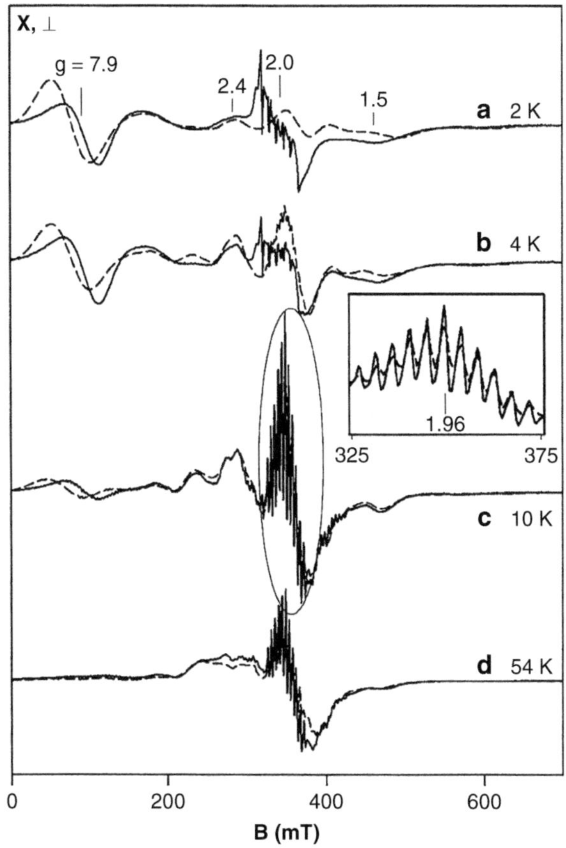

Fig. 3.9 Variable temperature, perpendicular mode EPR spectra (*solid lines*) and simulations (*dashed lines*) of an 11.5 mM sample of Mn dimer in 50:50 CH_3CN:DMF. From Golumbek and Hendrich (2003)

mode spectra at X-band and Q-band as a function of temperature are simulated simultaneously (Golumbek and Hendrich 2003). In this spin system S = 0 (ground state), S = 1, S = 2 etc. states contribute to different extents as the temperature is increased. This chapter emphasizes the quantitative analysis of dinuclear Mn(II) complexes by using simulation of the spectra.

The work of Hendricks and coworkers contains other many important examples of spectra obtained with $B_1 \parallel B_0$. Quantitative analysis of integer spin systems was discussed in Juarez-Garcia et al. (1991). A coupled Fe(III)–Cu(II) complex was compared with simulations and with a standard consisting of a single crystal of

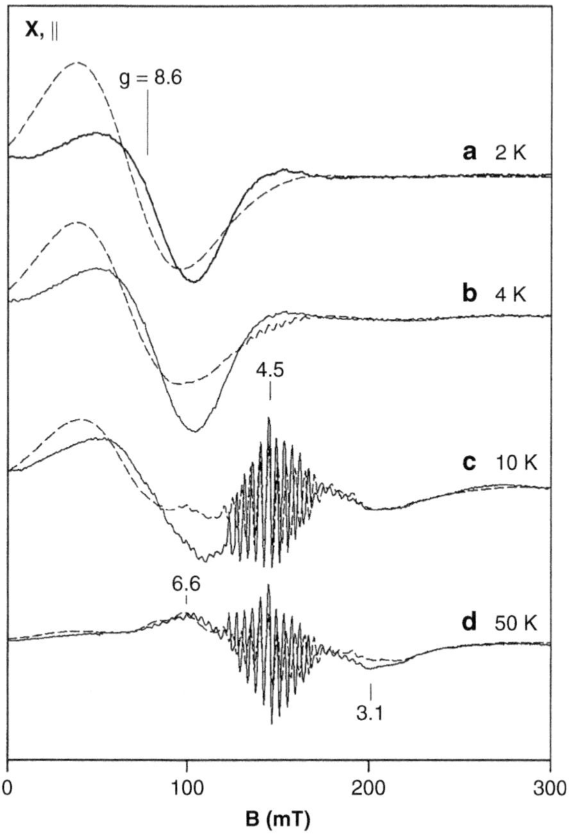

Fig. 3.10 The spectra were obtained with an X-band Bruker ER 4116DM resonator in ⊥ and || mode as a function of temperature. The paper also reports ⊥ and || mode Q-band spectra. From Golumbek and Hendrich (2003)

Fe(II)-doped zinc fluorosilicate. Hendrich and DeBrunner (1989) found that aquo Fe(II) was not a good reference standard for integer spin quantitation because the fraction of total spins observed is small, and the spectrum depends on sample preparation.

Chapter 4
A More in Depth Look at the EPR Signal Response

Optimizing an EPR experiment depends on sample preparation, an understanding of the interaction of the sample with the spectrometer, and the selection of appropriate operating parameters.

4.1 Sample Preparation

Quantitative EPR ultimately depends on the use of an analytical balance and volumetric glassware. Since concentrations of samples are low, and limited quantities of sample may require preparation of small volumes of solution, considerable care is needed to weigh and dissolve samples accurately such that this step does not limit the ultimate accuracy of the EPR measurement.

Solvent selection is critical for fluid solution studies because of the effect of a sample's dielectric loss on the resonator Q (see Chap. 7). Because microwave dielectric loss is due to molecular motion, which decreases upon freezing, the impact of solvent on resonator Q is much smaller when samples are frozen. Note, for quantitative work, however, that sample volume changes when the temperature changes, so the concentration in terms of spins per cm^3 of sample changes upon freezing.

4.1.1 Capillary Tube Sealant

When aqueous solutions are studied using capillary tubing, there are several techniques for sealing the tubes. If the sample must be maintained under a defined atmosphere, then after preparing the sample in the appropriate condition, the tube is flame sealed. If the goal is to maintain an atmosphere of defined composition, but the sample can be handled in air, then the sample can be put into a thin-wall Teflon tube or a machined TPX sample tube (these are fragile and expensive, but very

G.R. Eaton et al., *Quantitative EPR*,
DOI 10.1007/978-3-211-92948-3_4, © Springer-Verlag/Wien 2010

effective). The Teflon tube can be "closed" at the bottom by simply folding it. If the sample is in glass or quartz capillary tubes (round or rectangular) the bottom of the tube can be closed with Critoseal, which is sold by major lab suppliers. However, Critoseal contains a clay material that has a fairly strong Mn(II) EPR signal. The Critoseal plug has to be outside the active region of the resonator to avoid possible interference with the signal under study. The Bruker X-Sealant is a grease that was developed specifically to avoid the Mn(II) contamination problem when the sealed portion of the tube is placed in the active region of the resonator.

4.2 Searching for a Signal (Also See Appendix A)

When initially searching for a signal in a sample whose spectroscopic properties are not known, one can use relatively high spectrometer settings. For example, 10 mW power, 1 Gauss modulation amplitude, a fast scan, and short filter time constant are often adequate for detecting an unknown organic free radical species. These settings usually allow one to at least detect a signal if it is present in a reasonable concentration and detectable at the temperature selected. Recognize, though, that such a cursory scan could miss samples at two extremes: (a) a signal with such long relaxation time and narrow line that it is saturated or filtered out, or (b) a broad signal in the presence of a more obvious sharp signal. Always look for spectra that are not expected. A quantitatively correct spectrum requires adjustment of microwave power, phase, modulation amplitude, receiver gain, scan rate, and filter time constant. The criteria for selection of these settings are discussed in the following paragraphs.

4.3 Detector Current

The output of the crystal detector in the microwave bridge depends upon the magnitude of the bias current to the crystal detector. Each spectrometer should be checked to determine the range of detector current values within which the signal amplitude is independent of detector current. If the detector current drifts, as can happen with lossy solvents, or when the temperature is changed, significant errors in signal amplitude can result; S/N is degraded, and quantitative measurements are prevented. The output of the detector crystal is dependent on temperature. Since the bridge warms up during the first hour or so after power is turned on, the accuracy of quantitative spectra may change during this period. Similarly, the detector crystal itself changes temperature as a result of changes in incident power, so a spectrum acquired immediately after a large microwave power change may not provide an equilibrium response. In the extreme case of very strong signals, the detector current may vary as the field is scanned through resonance. If this occurs, then the sample size should be decreased.

4.4 Optimize the Receiver Gain

The gain is adjusted to give the desired size of display and where possible should be increased to use the full range of the digitizer. The operator must always check for linearity if blessed with a signal strong enough to saturate some component in the detection system. One aspect of signal averaging for very noisy spectra (S/N < 1) is sometimes overlooked: If the noise is being clipped, the signal is also being clipped. The maximum excursions of the noise must be within the range of the digitizer. In newer spectrometers 24 bit digitizers are used, so that the detector has a very high dynamic range.

The receiver gain must be high enough to show all details in the spectrum. Figure 4.1 shows the results of insufficient or excessive receiver gain. If the receiver gain is too low the effect of digitization will be evident in the spectrum (spectrum b), whereas at high gain the signals will be clipped due to an overload in the signal channel (spectrum c).

4.5 Be Aware of Noise Sources

Recognition of noise sources is useful in optimizing parameters. Noise in an EPR spectrometer may not be "white noise" (uniform noise amplitude over all frequencies). Non-white noise can occur due to various instabilities, ranging from electrical

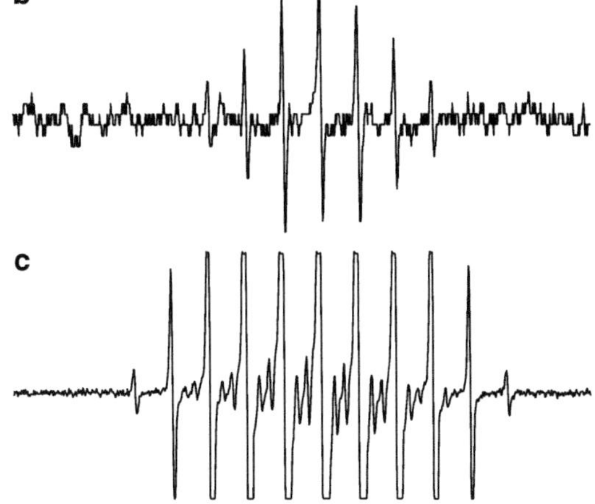

Fig. 4.1 Effect of using gain settings that are either (**a**) optimal, (**b**) too low, or (**c**) too high on an EPR spectrum

power instabilities to building vibrations. There usually is some 1/f noise (noise that is higher at lower frequency), so there is a tendency for better S/N at higher modulation frequencies. It is possible that particular modulation frequencies will give better S/N than other frequencies. White noise is also called random noise, and upon co-adding spectra, random noise decreases relative to signal, proportional to the square root of the number of scans averaged. Some "noise" is determinant, that is, it always occurs at particular frequencies – it is unwanted signal due to something other than the EPR signal. Usually, these unwanted signals can be removed by subtracting a background. However, background subtraction also increases random noise by $\sqrt{2}$ (unless the background is "smoothed" before subtraction).

4.6 Number of Data Points

Modern computer controlled spectrometers have as many as 256,000 magnetic field data points possible at 24 bit resolution. The optimal minimum data point selection is essentially the same as the calculation for time constant; there should be at least ten magnetic field data points over the narrowest line in the spectrum. Thus, if the line is 0.1 G wide, magnetic field steps should be no larger than 0.01 G, which with 8,192 magnetic field steps limits the spectral scan width to ca. 80 G. Figure 4.2 shows the enhancement in resolution that is achieved by using an eight-fold increase in the number of data points. Such high resolution in the magnetic field is very important when trying to resolve spectra with many narrow lines or features.

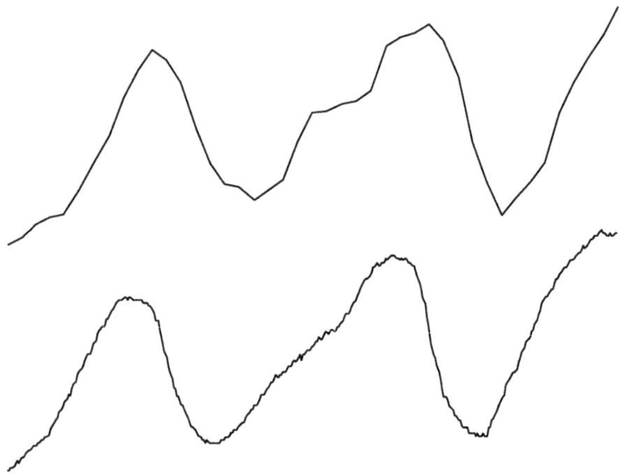

Fig. 4.2 An expanded view of an EPR spectrum using 1,024 data points (*upper trace*) or 8,192 data points (*lower trace*)

4.7 Optimize the Sweep Time and Conversion Time

Decisions concerning sweep time and filtering are correlated, and sample dependent. For a strong signal it may be possible to achieve adequate signal-to-noise with a single scan, modest filtering, and a relatively fast scan. However for weaker samples decisions need to be made about trade-offs between filtering, scan rate, and signal averaging.

Signal-to-noise can be improved by using a long time constant, slower scans, and fewer averages; or by using a small time constant, faster scans, and more signal averaging. Why use multiple faster scans instead of using a very long scan with extensive analog filtering? The answer is that with a perfectly stable sample and stable instrument, roughly equal time is involved in either method of S/N improvement. The problem is that perfect stability is not achieved, and the filtering discussion focuses on high frequency noise. Long-term spectrometer drift due to air temperature changes, drafts, vibration, line voltage fluctuations, etc., limit the practical lengths of a scan. Signal averaging tends to average out baseline drift problems along with high frequency noise. Drifts in the magnetic field magnitude are not averaged out by filtering or slow scans, and always increase apparent line width. Ultimately, the line broadening caused by magnetic field deviation limits the spectral improvement that is possible with any signal averaging or filtering technique. In addition, computer collected spectra can subsequently be digitally filtered without changing the original data, whereas analog-filtered data are irreversibly modified.

If the sample decays with time, a separate set of problems emerges. Assume, for example, that the goal is to compare line shapes of two peaks in a noisy nitroxide EPR spectrum, and that the amplitude of the spectrum is changing with time due to chemical reaction (shifting equilibria, decay, oxygen consumption or diffusion, etc.). In this case one wants to minimize the time spent scanning between points of interest. It would be wise to scan the narrowest portion of the spectrum that will give the information of interest. Then a numerical correction for the measured rate of change in the spectrum is the best way to handle the problem. The impact of the time dependence can be minimized more effectively by averaging rapid scans than by filtering a slow scan.

In older spectrometers the sweep time was an independent parameter. In newer spectrometer the sweep time is digitally controlled and the sweep time is determined by the dwell time at each field step. For Bruker spectrometers an integrating digitizer is used and the integration time of the digitizer is the "conversion time," which is also the magnetic field dwell time. The integration time is the time during which the analog to digital converter (ADC) accumulates the signal and noise at each magnetic field step of the EPR experiment. The time for a magnetic field scan is the product of the conversion time (in seconds) and the number of data points acquired on the magnetic field axis. For example, a conversion time of 81.92 ms for 2,048 magnetic field steps results in a scan time of 167.8 s. For the Bruker digitizer the conversion time affects the dynamic range of the experiments. To resolve lines that are very intense as well as lines that are very weak (for example, carbon 13 satellites) within the same spectrum a sufficiently long conversion time is needed.

4.8 Optimize the Time Constant for the Selected Sweep Time and Conversion Time

The decrease in the standard deviation of random (white) noise is proportional to the square root of the time constant. Thus, if the filter time constant is increased by a factor of 4 from 20.48 to 81.92 ms, the noise will be reduced by a factor of 2. When the time constant is longer than the conversion time, the S/N improvement is proportional to the square root of the filter time constant.

If a time constant is selected that is excessively high relative to the sweep time, the signal may be filtered out! The time constant should be adjusted to be consistent with the A/D conversion time. The time constant should be sufficiently long to filter out undesirable noise, yet short enough that it does not distort the signal. Therefore, if a longer time constant is needed, the scan time (conversion time) should be increased. Figure 4.3 shows the effect of progressively increasing the time constant while maintaining the same sweep time. All the spectra are at the same scale.

Although the S/N increases when the time constant is increased, the spectrum is distorted if the time constant is too long (see Fig. 4.3d). As a "rule of thumb", the time constant should be chosen to be less than about 1/10 the time it takes to scan

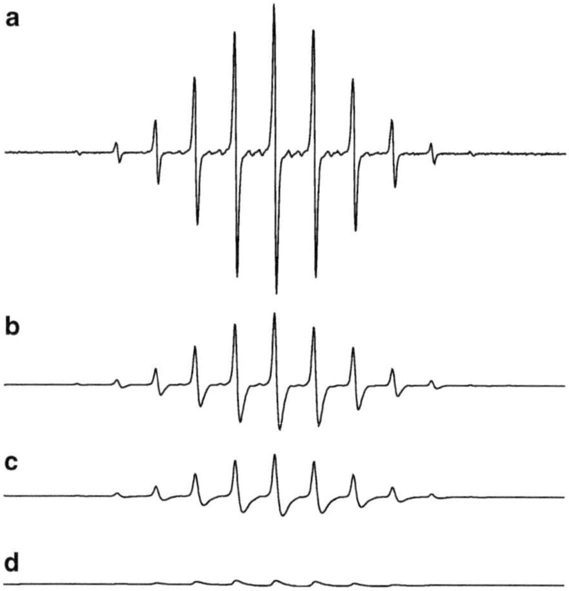

Fig. 4.3 Effect of using a progressively longer time constant (**a–d**) on an EPR spectrum

through the narrowest line in the spectrum. Consequently, scan rate and filter time constant are related to each other and to the CW line width in the following formula.

$$\frac{(\text{spectrum width in Gauss})}{(\text{line width in Gauss})} \times \frac{(\text{time constant in sec})}{(\text{sweep time in sec})} < 0.1 \qquad (4.1)$$

The inequality must be satisfied to obtain undistorted lines. A faster sweep or longer time constant does not give the system enough time to respond to changes in signal amplitude as the line is traversed.

Example. To record a 20 G scan of a 0.1 G wide line in 84 s, the time constant should be less than 0.04 s; (20 × 0.04)/(0.1 × 84) = 0.095. If a 40.96 ms conversion time is used, the scan will be 84 s if 2,048 steps are used. 1,024 steps would result in a scan of 41 s, which would not be conservative with respect to line shape (and position) distortion.

Another feature of many integrating digitizers is that the signal level increases the longer the integration time (conversion time). Thus, if two spectra of the same sample are obtained with all parameters equal, but one with 10.24 ms conversion time and one with 81.92 ms conversion time, the numerical integral of the digitized spectrum will be eight times larger for the 81.92 ms conversion time. In Bruker spectrometers running the Xepr acquisition software, these are the numbers that are displayed in the Xepr window, are saved as the amplitude of the spectrum, and will affect integrals of the area under the peaks, etc. Since this could cause undesirable confusion when comparing spectra, some users prefer to have spectra displayed and stored numerically normalized. This is often an option in the software and in some new spectrometers is performed automatically.

4.9 Background Signals

"There are spins everywhere". (James Hyde, personal communication to almost everyone.) There are EPR signals even if there is no sample in the resonator. Actually, the signal from O_2 in room air is easily measured at Q-band, and was used by Varian to calibrate sensitivity of Q-band spectrometers. There are EPR signals from metal ions in both metallic and dielectric resonators. There are impurity and defect signals in quartz sample tubes and Dewar inserts, even when these are made of the highest purity quartz available. (Pyrex tubes yield an enormous iron signal.) Dirt transferred to the outside of sample tubes by routine handling will also yield EPR signals. Even some "pure" Teflon has a fairly strong EPR signal.

The EPR signals that cannot be attributed to the sample of interest are generally called "background" signals, whether they are due to unwanted spins or to electro-mechanical effects on the resonator. Background signals should be recorded and subtracted from the spectrum of the sample of interest to obtain quantitative results.

4.10 Integration

The area under the absorption curve (obtained from double integration) is needed to precisely quantitate the intensity. However, in cases where the same species is present in a series of samples and the peak-to-peak linewidth is invariant, comparison of peak-to-peak amplitude is a convenient method of quantitating signals. Even if the linewidth changes, if the line shape (i.e. % Loentzian) does not change from sample to sample, it is valid to approximate the relative areas using the peak-to-peak width squared multiplied by the peak-to-peak amplitude (Chesnut 1977). These are useful alternatives when a noisy drifting or otherwise confounding baseline prevents an accurate double integral calculation.

The area under a derivative EPR signal is obtained by digital double integration. It is now so common to integrate spectra acquired into a computer, that the first and second integrals of an experimental derivative spectrum are commonly used to judge the quality (phasing, etc.) of the experimental data. Proper integration requires a constant baseline or a baseline fitting and correction routine to obtain a baseline that is constant on both sides of the EPR spectrum.

Significant error may occur in the double integration value when the field scan range of the experiment is not wide enough to encompass the entire EPR absorption envelope. The percentage error resulting from finite truncation of the curve is especially large for Lorentzian lines because of their extensive wings. For example, two linewidths from the center of a Gaussian or a Lorentzian first derivative line the signal amplitude is 0.2 or 17.2% of the peak amplitude, respectively. For the same integration accuracy, a Lorentzian line must be integrated over many more linewidths than a Gaussian line. In practice, the wings of a signal tend to disappear into the noise so one might be tempted to select a narrower scan than is needed for accurate integration. Figure 4.4 shows the error that is caused in the integral by

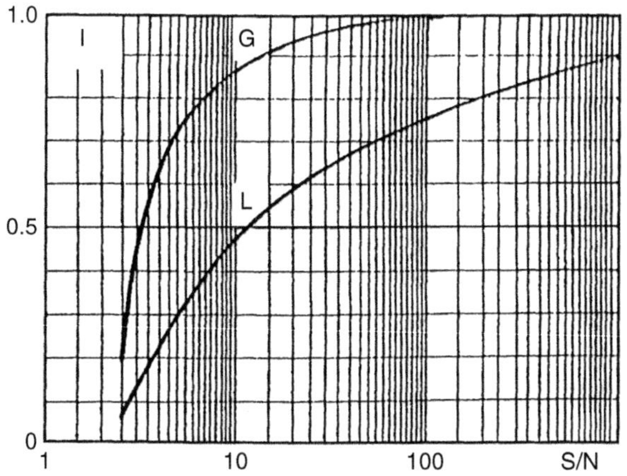

Fig. 4.4 Integrated area under truncated Lorentzian (L) or Gaussian (G) lineshape as a function of signal-to-noise ratio. From Czoch (1996)

truncating the spectrum at the point with the signal amplitude equals the noise. Very high S/N is needed to use this criterion for selecting scan width.

Beyond these matters, there are spectra that are inherently difficult to integrate, because they do not meet the ideal case described above of isolated lines with well-defined flat baselines on either side of the EPR spectrum. The spectrum of interest might, for example, be a semiquinone radical superimposed on other unknown species, or the spectrum might be so wide that it is not possible to scan the field far enough to define the entire baseline. If the S/N is very low, if there is an overlapping spectrum from another species, or the background is large and unavoidable, it may be best to integrate a computer simulated spectrum by fitting the experimental data. For example, one could simulate the full spectrum using only parameters that fit the region of interest. Performing a double integration on this simulated spectrum would only reveal the EPR area under the portion of interest. If any prior knowledge of the spectrum can be used, the accuracy of the results can be improved. For example, Halpern and coworkers have used spectral fitting to enhance quantitative EPR imaging (Halpern et al. 1993, 1994; Mailer et al. 2003).

4.11 Microwave Power

If quantitating spin concentration is the goal of the experiment, it is necessary to acquire the spectrum at a microwave power level well below that where significant saturation of the EPR spectrum occurs. Most EPR samples can be saturated with the power levels available in commercial spectrometers. Thus, it is always important to check for saturation.

In the absence of saturation, the EPR signal increases linearly with B_1. The efficiency of the resonator to actually convert microwave power to B_1 is an important factor and can vary widely between different resonator designs. The Q-factor and filling factor of the resonator are also important. These depend on the type of resonator and the dielectric properties of the sample (this is discussed in more detail in Chaps. 6–8).

If there is no unresolved hyperfine splitting, the relationship between linewidth, B_1, and relaxation times is

$$\left(\Delta B_{pp}\right)^2 = \frac{4}{3\gamma^2 T_2^2}\left(1 + \gamma^2 B_1^2 T_1 T_2\right), \tag{4.2}$$

where B_{pp} = peak-to-peak line width and γ = electron magnetogyric ratio (Schreurs and Fraenkel 1961; Eastman et al. 1969; Poole and Farach 1971). If the B_1 field is small enough that the $\gamma^2 B_1{}^2 T_1 T_2$ product term is <<1, then the signal is unsaturated. The equations for EPR line shape yield a term that is commonly called the saturation factor:

$$s = \frac{1}{1 + \gamma^2 B_1^2 T_1 T_2}. \tag{4.3}$$

In various texts the term "saturation factor" is used to mean the full expression in (4.3), the denominator of (4.3), or only the $\gamma^2 B_1^2 T_1 T_2$ term of the denominator. For microwave power levels below saturation, s = 1. A value of 0.98 or even 0.95 is an acceptable degree of saturation when acquiring CW EPR spectra.

It is useful for the operator to know for a given resonator, in general, the relationship between B_1 at the sample and the incident microwave power. Often, the first experiment that should be performed with a new unknown sample is a power saturation study. This will give the operator a good idea of the power at which the sample starts to saturate, and what power provides optimal S/N with minimal saturation.

One has to be careful comparing power saturation at different microwave frequencies (e.g., X-band vs. Q-band) because of the use of different resonators. The effect of a dielectric, such as a sample, sample tube, or especially a VT Dewar, on the distribution of B_1 in a resonator is sometimes called the "sucking in effect" in the EPR literature. Modern electromagnetic field calculations, such as with the Ansoft Corporation computer program HFSS (high frequency structure simulator) reveal the details of microwave field distributions, as shown by the recent papers from the Hyde laboratory about aqueous sample cells and uniform field resonators (Mett et al. 2001; Anderson et al. 2002; Hyde and Mett 2002; Hyde et al. 2002; Mett and Hyde 2003; Sidabras et al. 2005).

A series of measurements at different power levels is called a progressive saturation study. A plot of such results will look like one of the lines in Fig. 4.5.

The incident microwave power should be set to a value below that at which the power saturation curve deviates from linearity. To obtain quantitative EPR results (particularly for measuring spin concentration) the operator needs to check that the spectrum is measured in the linear B_1 range. In the linear range the EPR signal amplitude increases proportional to the square root of the microwave power. A simple way to check the linear power range for a given sample is to increase the power by a factor of 4 (i.e., decrease the attenuation by 6 dB). The spectral amplitude should increase by a factor of 2; if it doesn't, reduce the power and try again. Since there is no strictly linear range, except for species with very short relaxation times, the operator has to make a judgment about how much deviation from linearity is acceptable. A good rule of thumb is to use a microwave power level that provides minimal saturation yet, is high enough that uncertainty in the measurement due to low S/N is not a problem.

The power saturation plots of a crystalline BDPA sample are shown in Fig. 4.6. The CW line width of ca. 0.7 G corresponds to $T_2 \approx 1 \times 10^{-7}$ s, so the line is relaxation determined (calculations described in More et al. 1984).

BDPA is a good sample for estimating the B_1 field in a resonator by CW EPR. To get a good estimate of B_1, it is necessary to use power levels that go beyond the point where the curve deviates from linearity. This was an important criterion in selecting the BDPA sample. If one uses too much power, the EPR lines are broadened, as shown in Fig. 4.7.

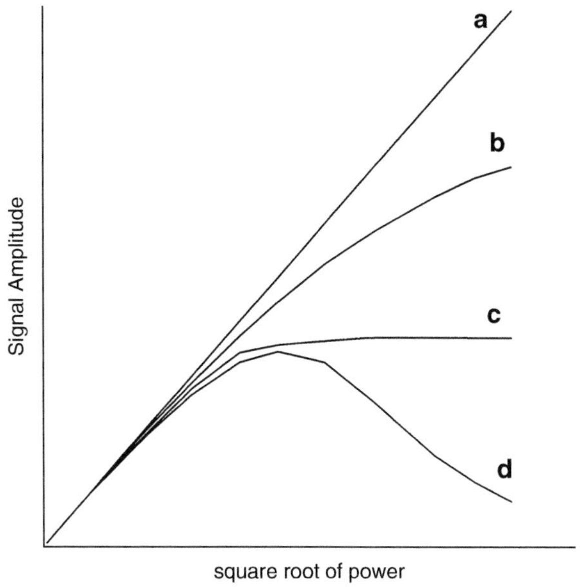

Fig. 4.5 A plot of signal amplitude as a function of the square root of the microwave power incident on the cavity is called a power saturation curve. The figure presents CW EPR power saturation curves for four cases: (**a**) no power saturation within the available microwave power of the spectrometer – short relaxation times; (**b**) a small degree of homogeneous saturation at the highest available power levels – intermediate relaxation times; (**c**) extensive inhomogeneous saturation, exhibiting a "leveling" of the saturation curve at relatively low power, but no subsequent decrease in amplitude of the EPR signal; (**d**) extensive homogeneous saturation, exhibiting a maximum in the saturation curve, due to long relaxation times. If only the segment of the curve shown in (**b**) were observed, it would not be possible to distinguish between homogeneous and inhomogeneous broadening. From Eaton and Eaton (2005)

Even as pulsed EPR spectrometers become more common, CW progressive power saturation studies of spin relaxation will remain important. Species with shorter relaxation times can be better characterized by CW methods than by pulse methods. In addition, the methods are complementary in the information they provide. In the literature $P_{1/2}$ is commonly used as a saturation parameter (Mailer et al. 1977). $P_{1/2}$ is the incident microwave power at which the EPR signal has half the amplitude it would have in the absence of saturation.

The estimation of B_1 using a point sample such as a crystal of BDPA positioned in the center of the resonator defines the maximum value. It is inherent in a resonator that the microwave field depends on position. The B_1 field in a TE_{102} cavity varies sinusoidally along the length of the tube and therefore saturation varies along the sample. Because of the distribution in B_1 and modulation amplitude, the most intense signal is from the center of the resonator. If that segment of the sample is not saturated, then other portions of the sample are also not saturated.

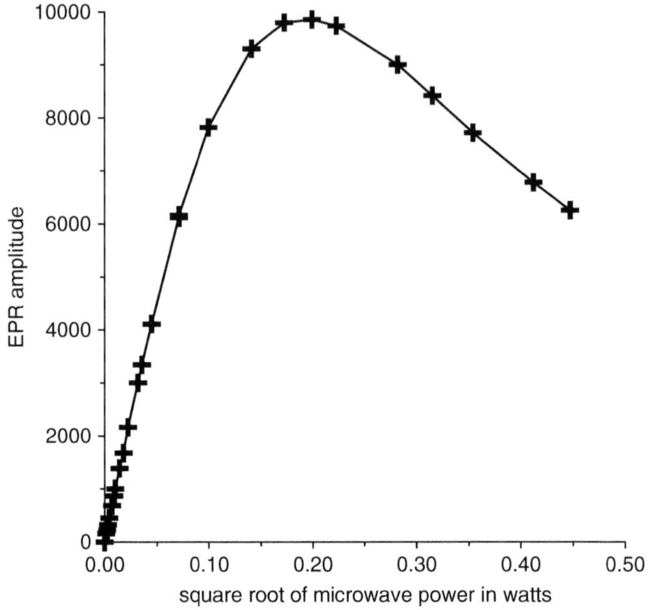

Fig. 4.6 CW power saturation curve for BDPA in a TE_{102} cavity

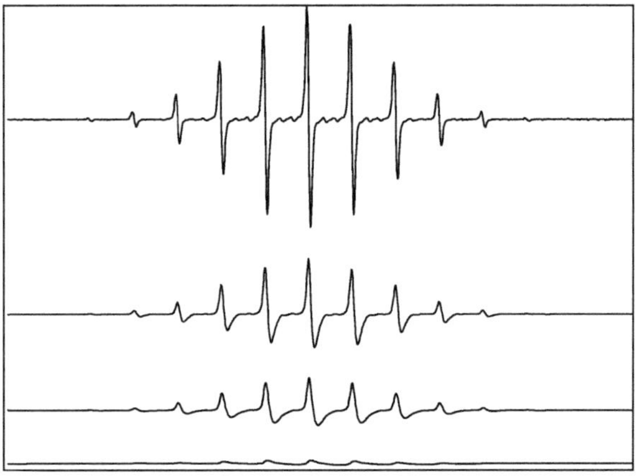

Fig. 4.7 Increasing microwave power (*top* to *bottom* in this figure) broadens the EPR lines

4.12 Modulation Amplitude (Also See Appendix B for More Details on This Topic)

If a sample is strong enough, direct detection from the crystal detector would reveal an absorption signal as sketched in Fig. 4.8. Unfortunately, the S/N for most samples is not adequate for this method of direct detection. In fact, the EPR signal

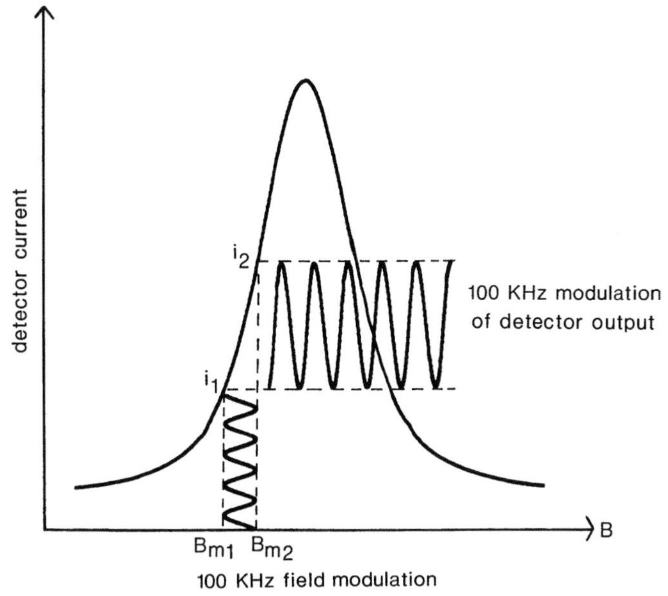

Fig. 4.8 The principle of magnetic field modulation and phase-sensitive detection at the modulation frequency is illustrated by this sketch. The EPR absorption curve is traced out as the magnetic field is swept at constant microwave frequency. From Eaton and Eaton (2005)

intensity is rarely larger than the noise until some method is used to separate the signal from the noise. In CW EPR, as in many other fields of science, the S/N is increased by encoding the desired signal at a particular frequency. This is done by modulating some parameter that affects the signal (at this particular "modulation" frequency), and then separating the desired signal from noise or other incoherent undesired signals using phase-sensitive detection at the modulation frequency. A lock-in amplifier selects only signals at the modulation frequency and filters the signal of interest with a selectable time constant. The result is that noise is suppressed and the S/N of the EPR signal is greatly increased.

Figure 4.8 shows the principle of magnetic field modulation and phase-sensitive detection. The modulation amplitude should be small enough that it samples an approximately linear segment of the EPR absorption signal. Of course, no portion of the absorption signal is actually linear, and the importance of the degree of distortion from a true derivative is a matter of operator judgment. With regard to quantitative EPR, the choice of modulation frequency and modulation amplitude can be critical. The operator must decide what information about the spin system is desired. If the details of line shape are important, then the spectrum has to be obtained under conservative conditions that do not distort the spectrum.

Wilson (1963) calculated the effect of modulation on the shape and area of Lorentzian and Gaussian lines, and showed that derivative line shapes occur when the modulation is sufficiently small. See also the papers by Burgess and Brown

(1952), Wahlquist (1961) and Smith (1964) and the extensive discussion in Poole (1967, 1983). Figure 4.8 shows how the derivative line shape occurs in the phase-sensitive detection when the modulation is small. Small, in this case, means that the amplitude of the modulation in the magnetic field direction samples a portion of the absorption curve that is locally linear. The amplitude and the phase of the detected signal (labeled as at the common modulation frequency of 100 kHz in the figure) is positive on the low-field side of the line, is zero at the peak of the line, and then is negative on the high-field side of the line. The magnitude of the EPR signal is related to the amplitude of the magnetic field modulation. To the extent that the portion of the line sampled is nearly linear, the amplitude of the signal increases with increase in the modulation amplitude. However, this linear response disappears when the amplitude becomes large enough that the curvature of the Lorentzian or Gaussian (or more complicated) line becomes significant. Changes in the lineshape that occur with increasing modulation amplitude are shown in the following section, which describes the modulation amplitude calibration procedure.

When the modulation amplitude ($B_{m2} - B_{m1}$ in Fig. 4.8) is less than about 1/10 of the derivative peak-to-peak line width, ΔB_{pp}, the distortion of the signal is only a few percent. Increased signal amplitude comes at the cost of increasing distortion of the EPR line shape. However, modulation amplitudes up to about 1/3 of ΔB_{pp} cause relatively little distortion. For the most accurate lineshape, the spectrum should be acquired with several very small modulation amplitude settings. When decreasing the modulation amplitude causes no further apparent decrease in linewidth, set the modulation amplitude to 1/10 of that linewidth. For small modulation amplitudes the S/N increases linearly with modulation amplitude. For some samples the S/N might be too low to obtain a spectrum with the modulation amplitude meeting the above (1/10 linewidth) criteria. In this case, one has to go back to the basic question – what information is wanted from the sample? If lineshape is the crucial information, then longer acquisition times through the use of signal averaging will be required to improve the S/N. If lineshape is of less significance, it may be acceptable to increase the modulation amplitude up to roughly 1/2 of the linewidth. If the area under the peak is the information desired, overmodulation is acceptable, since the area is linearly proportional to modulation amplitude, even when the modulation is so large as to cause distortion of the lineshape. Computational approaches have been developed to correct for the lineshape distortion due to overmodulation (Robinson et al. 1999a, b). In more complex spectra where there are several overlapping lines, the modulation amplitude must be selected based on the narrowest feature to be resolved. Hyperfine structure can be obscured by using excess modulation amplitude.

For wide EPR lines, as occur for many metal ions, especially in rigid lattice, it often is not possible to use modulation amplitude as large as 1/10 of the line width. The modulated magnetic field produces eddy currents in metallic parts of the resonator, and since this is in a magnetic field, vibrations can occur. For each resonator there will be a modulation amplitude level at which microphonics will occur adding noise to the spectrum. Thus, the gain in signal amplitude from exceeding this level no longer benefits the overall S/N. Furthermore, eddy currents result in heating of the resonator, which is also undesirable. Spectrometer manufacturers specify a modulation amplitude maximum for each type of resonator.

4.13 Modulation Amplitude Calibration

To perform quantitative EPR comparing samples run in different laboratories or even in different resonators in the same lab, it is necessary to calibrate the modulation amplitude. Calibration assures that the modulation amplitude level selected in the software (or selected by a front panel switch in older spectrometers) is equal to what the hardware is actually producing at the sample.

The distortion of the EPR line by modulation amplitude that is large relative to line width (Fig. 4.9) can be used to calibrate the modulation field. As shown in Poole (1967, pp. 398–413) and (Weil et al. 1994b, pp. 554–556), when the modulation amplitude is about 1/3 of the line width, a Lorentzian spectral line is broadened by about 3%, and when the modulation is ca. 2/3 of the line width, the line is broadened by about 11%. How important this broadening is depends on the information one desires from the spectrum. Many tables and plots of the effects of modulation on line widths, shapes, and areas are given in Poole (1967, Chap. 10). This tabular data reveals, for example, that if a line is 1 G wide and the modulation amplitude is 4 G, the observed overmodulated line will be ca. 3.5 G wide.

The spectrometer's signal channel reference phase and modulation amplitude need to be carefully calibrated to obtain maximum sensitivity, minimum distortion, and quantitatively reproducible measurements. In a typical automated calibration routine, the results of the calibration are saved on disk for future use. It is recommended that a recalibration be performed at least once a year to ensure quantitative and reproducible results. Each cavity or resonator has its own individual calibration. Therefore, this procedure must be followed for each cavity.

Calibration of the signal channel involves two separate yet interdependent procedures. The first procedure is to calibrate the peak-to-peak modulation amplitude. For the sake of brevity, modulation amplitude will be used in place of peak-to-peak modulation amplitude. The second procedure is to calibrate the phase difference

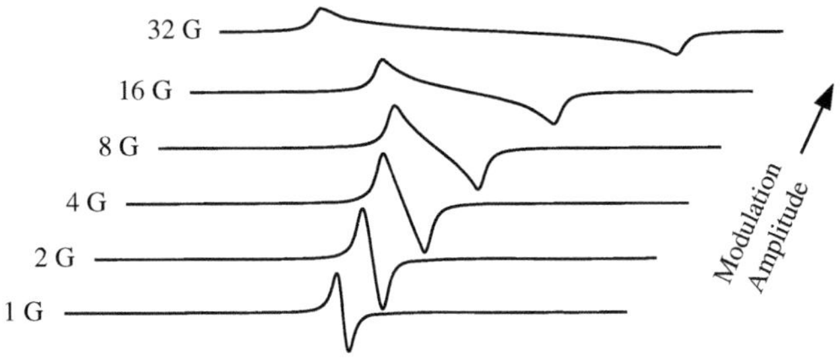

Fig. 4.9 The signal shape of the DPPH EPR signal as a function of increasing modulation amplitude

between the reference signal and the modulated EPR signal. Because the calibration and adjustment of the modulation amplitude can affect the phase difference, the first procedure is performed first.

The modulation amplitude is calibrated by overmodulating a narrow EPR signal. A crystal of DPPH or BDPA, with a line width of approximately 1 G, is a very good sample to use. The sample should be placed accurately at the center of the resonator. When the modulation amplitude is large compared to the line width, the magnetic field modulation brings the sample into resonance before and after the magnet has reached the field for resonance. This results in a broadening and distortion of the EPR signal (Fig. 4.9). In the limit of an infinitesimally narrow EPR signal, the peak-to-peak width of the first-derivative EPR signal is approximately equal to the peak-to-peak modulation amplitude.

The first step in calibrating the modulation amplitude involves choosing the correct tuning capacitors (Fig. 4.10). The voltage required to produce the current that results in the modulation field is proportional to the modulation frequency. At frequencies greater than about 50 kHz, the voltage required to produce the current would be beyond the range of the amplifiers used in most spectrometers. When a series resonating capacitor is used, and the frequency is at the circuit resonance, the amplifier only has to produce a voltage that is proportional to the amplitude of the current, not the time derivative of the current. Thus, the calibration of the modulation amplitude involves selecting the capacitors to resonate (also called tune) the modulation coil to keep the voltage low at high modulation frequencies (Fig. 4.10).

The calibration routine switches various tuning capacitors in and out of the circuit until the modulation amplitude is maximized. The optimal capacitor for that particular frequency as well as the modulation amplitude for full gain of the modulation amplifier are recorded and saved with the calibration file. Once this data is available, the signal channel will then vary the input signal to the modulation amplifier to produce the selected modulation amplitude.

Once the modulation amplitude has been calibrated, the reference phase is easily calibrated by studying the phase angle dependence of the signal intensity. The intensity of the output signal is proportional to the cosine of the phase difference between the reference signal and the modulated EPR signal (Fig. 4.11). It is most

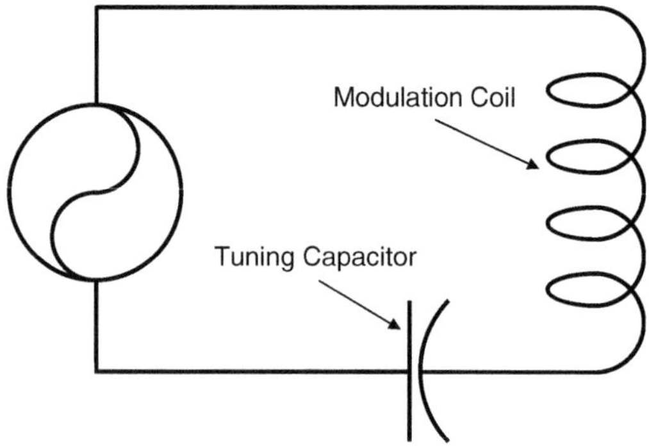

Fig. 4.10 The LC resonant circuit for high frequencies

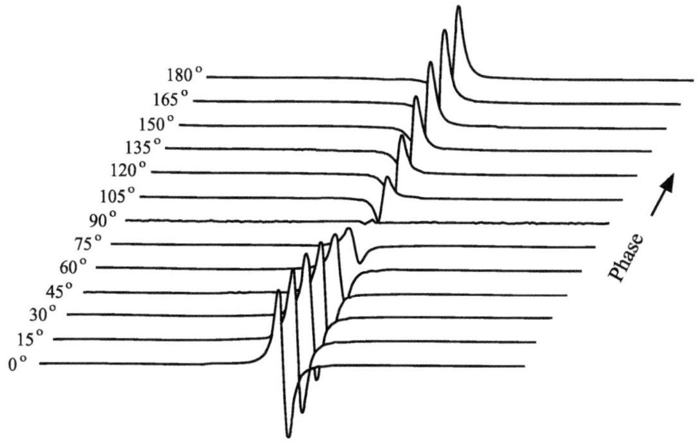

Fig. 4.11 Signal intensity as a function of the reference phase angle for modulation

convenient to determine where the 90° phase difference occurs because first, the absence of a signal (cos90° = 0) is easy to detect and second, the cosine function (and hence the intensity) changes rapidly with respect to the phase angle at 90°. In the calibration routine, spectra are acquired at several different values of the reference phase and the 90° phase difference is extrapolated from the signal intensities. The phase angle resulting in maximum signal intensity for that particular frequency is recorded and saved with the calibration file.

The phase difference between the modulated EPR signal and the reference signal depends on several experimental conditions. The length of the cable leading to the modulation coils, the inductance of the coils in the particular cavity, the gain setting of the modulation amplifier, the tuning capacitors, and the signal channel used can all change the phase difference. The reference phase calibration is performed automatically during the Bruker software routine described in this section. For older spectrometers these calibrations can be performed manually.

The two editions of the book by C.P. Poole that are mentioned at the beginning of this chapter are very good references for the details on the theory of phase-sensitive detection and the calibration of signal channels. The reader is encouraged to explore this topic further to learn more about calibration.

This calibration is for a point sample at the center of the resonator. The variation of modulation field over an extended sample depends on the construction of the modulation coils and resonator. In a standard TE_{102} cavity the modulation field variation is approximately sinusoidal with a maximum at the center and minima near the top and bottom. This variation is why it is important that the unknown and standard samples used in quantitation have similar geometry.

Although the modulation distribution is independent of a quartz Dewar insert, the B_1 distribution is strongly affected by the presence of a quartz Dewar insert in the resonator (and these inserts are not all the same). Therefore, calibrations of B_1 and modulation are specific to a particular Dewar and resonator combination.

4.14 How to Select Modulation Frequency

Due to the favorably low 1/f noise characteristics, most X-band EPR experiments use 100 kHz as the modulation frequency for the lock-in detection system. Some special cases arise where other modulation frequencies are preferable to 100 kHz. These include spectra where the sample has a relaxation time of 10 μs or longer (as 100 kHz corresponds to 10 μs). In addition, 100 kHz modulation corresponds to 35 mG which results in sidebands in the EPR spectrum. For lines that are hundreds of mG wide, these sidebands are not observed. However, for a sample with narrow lines (<35 mG) such as the trityl radical, LiPc (lithium pthalocyanine) or N@C_{60} the line shape can be distorted by the modulation sidebands (Fig. 4.12). In these cases, one will often use a lower modulation frequency. The example in Fig. 4.12 shows the effect that modulation frequencies of 100, 50 and 20 kHz have on a spectrum obtained from N@C_{60}. The modulation side bands are resolved at 100 kHz, they appear to broaden the EPR line at 50 kHz and are not observed in the spectrum at 20 kHz. Therefore 20 kHz is the appropriate modulation frequency for a sample such as this with very narrow lines.

At high modulation frequency and high modulation amplitude levels, the rate of passage through the EPR line may be faster than relaxation rates, which result in "passage effects" (see the following section). For example, a 1 G sinusoidal modulation at 100 kHz is a magnetic field scan of $\pi \times 10^5$ G/s at the point of maximum scan rate as shown in recent papers from the Eaton lab. Passage effects are not as unusual a phenomenon as might be expected. It is easily demonstrated by the standard irradiated quartz sample that is available from Wilmad (WGSR-01-4), and other narrow-line samples with long relaxation times. For slow-passage EPR, it

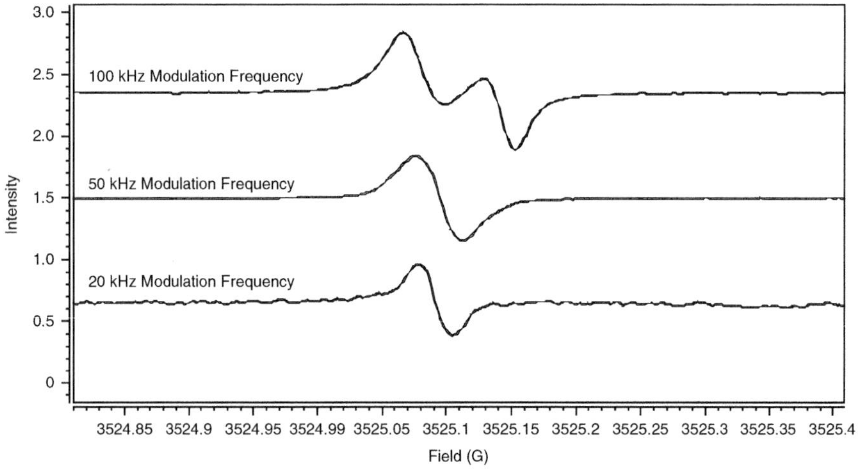

Fig. 4.12 N@C_{60} has a *very narrow line*. Modulation sidebands are observed when 100 kHz magnetic field modulation is used

is necessary to have the reciprocal of the modulation frequency much greater than T_1.

$$\nu_m{}^{-1} \gg T_1.$$

This criterion is not met as often as it is assumed to be. Some samples at liquid nitrogen temperature, and many samples at liquid helium temperature, have T_1 values that are too long to permit the use of 100 kHz modulation. Passage effects, recognizable as distortions of line shapes, or even inversion of signals upon reversal of the field scan direction, alert the spectroscopist to the need to use a lower modulation frequency. In the current Bruker spectrometers there is very little increase in noise at low modulation frequencies. So the use of lower modulation frequency does not compromise S/N.

Rapid scans can be utilized to provide additional information about samples (Stoner et al. 2004; Joshi et al. 2005b; Tseitlin et al. 2006).

4.15 Passage Effects

The previous section mentioned the occurrence of passage effects with certain samples when a 100 kHz modulation frequency is used. There are easily accessible experimental situations, especially at cryogenic temperatures, at which one observes these extreme relaxation-dependent phenomena (Weger 1960). When an EPR spectrum is scanned too fast to allow the spins to be at thermal equilibrium, passage effects occur. These passage effects result in distorted spectra that are best identified by recording spectra with both increasing and decreasing magnetic field scan rates, and by changing modulation amplitude and microwave power. In extreme cases, the first-derivative spectrum looks like an absorption spectrum, and the spectrum turns upside down when the field scan direction is reversed. An example is in Fig. 4.13. In the absence of passage effects the spectra from two field scans performed in reverse direction to one another should be super-imposable (assuming that the correct filter time constant has been selected.)

4.16 Illustration of the Effect of Modulation Amplitude, Modulation Frequency, and Microwave Power on the Spectra of Free Radicals

The accidental equivalence of hyperfine splitting and modulation sidebands can cause confusion about the number of hyperfine lines. This was illustrated by the change in apparent hyperfine in galvinoxyl radical (also called Coppinger's radical).

Spectra in Fig. 4.14 present two lessons. First, the resolution is better with 10 kHz modulation than with 100 kHz, as was also shown in Fig. 4.12. Second, increasing microwave power broadens the EPR lines due to saturation. An initial

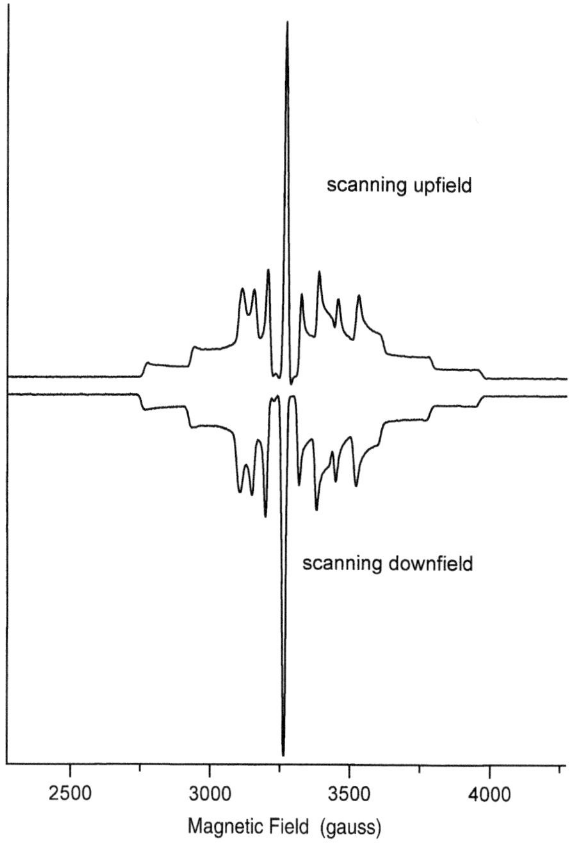

Fig. 4.13 X-band CW derivative EPR spectra of vanadyl porphyrin in frozen solution at ca. 5 K. From Eaton and Eaton (2005)

acquisition using 1.0 mW microwave power and 100 kHz modulation (which are good general starting parameters), would not reveal the additional hyperfine due to the t-butyl protons. In fact, the maximum spectral intensity occurs at 0.7 mW when 100 kHz modulation is used, but at 0.1 mW when 10 kHz modulation is used. Thus, the CW power saturation behavior is obscured by use of 100 kHz modulation when the lines are very narrow as in the galvinoxyl radical.

4.17 Phase

There are two phase settings in the normal CW spectrometer. Improper setting of either of these phase values can have detrimental effects on quantitative results. In each case, the wrong phase setting will result in an EPR spectrum with lower intensity than would be revealed under the proper phase setting. Thus, the amount

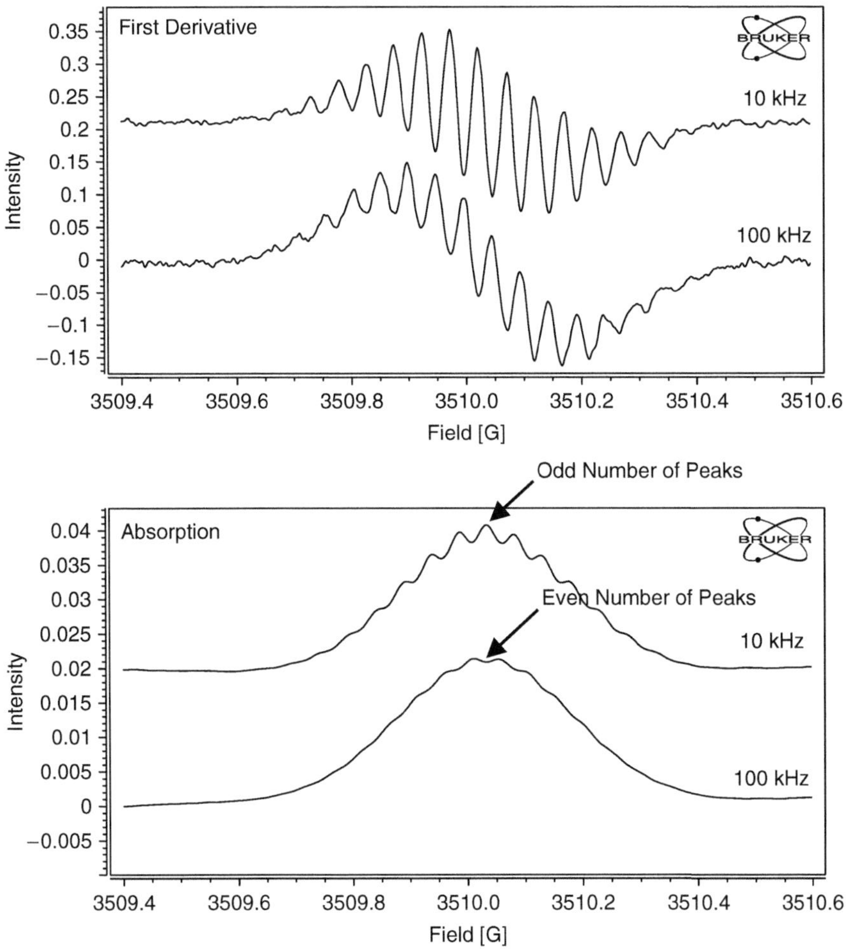

Fig. 4.14 A single line of the galvinoxyl radical, obtained with 100 and 10 kHz modulation and 16 mG modulation amplitude

of EPR signal (and hence the number of spins) will be underestimated. First is the reference arm phase in the microwave bridge. This phase setting is adjusted to maximize the detector current at high microwave power. If the reference arm phase is not correct, the signal amplitude is decreased and the dispersion signal will also be mixed with the absorption signal. The phase of the reference arm is optimized during the microwave bridge tuning procedure (which is automated in most modern spectrometers). The second phase is that of the reference signal that is mixed with the EPR signal in the phase-sensitive detection system. If the 100 kHz detector phase is wrong, the resulting signal coming from the mixer will also be low. A review of Fig. 4.11 shows the result of having reference signal phase at various values. To assure that the reference phase is correct, acquire a spectrum at 90°.

The spectrum should have very close to zero amplitude, as at 90° a properly phased reference signal should be exactly out of phase with the 100 kHz EPR. This results in a null signal. The reference signal phase is calibrated during the signal channel calibration routine, but is also configurable in the software. It should be set at zero for normal operation. For saturation transfer spectroscopy this setting is very critical, and a substantial literature has been devoted to it (Hyde 1978; Hyde and Thomas 1980; Watanabe et al. 1982; Beth et al. 1983). It is also important to adjust the phase of the AFC system (normally 70 or 77 kHz), but there is no user control of this phase in commercial instruments.

4.18 Automatic Frequency Control and Microwave Phase

The automatic frequency control (AFC) circuit is designed to keep the microwave source (klystron or Gunn diode) frequency locked to the resonant frequency of the resonator. The AFC is needed because the resonant frequency of the resonator can change due to temperature changes, changes in the sample, etc. The amplitude of the EPR signal has a cosinusoidal dependence on the frequency match, so it generally is not strongly affected by a slight change in the AFC correction signal. However, the AFC correction signal can mix in some dispersion signal with the absorption signal. If the sample is somewhat saturated, this mixing might be noticeable, because the dispersion signal does not saturate as readily as does the absorption signal. The standard irradiated fused silica sample (Eaton and Eaton 1993) sold by Wilmad as WGSR-01-4 has a long electron spin relaxation time T_1 and spin–spin relaxation time (see below), and the CW absorption spectrum mixes with the dispersion

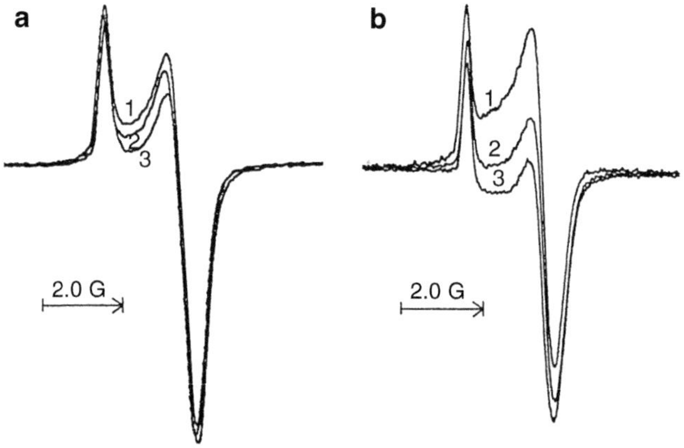

Fig. 4.15 Dispersion and absorption spectra mix when the AFC correction signal is large. The effect of this mixing can be seen sensitively when the electron spin relaxation time is so long that the absorption signal is partially saturated. Examples (**a**) and (**b**) are from two different spectrometers, and spectra 1, 2 and 3 are from microwave frequencies that differ only by a few megahertz. From Ludowise et al. (1991)

spectrum unless the frequency is very carefully set and controlled (Fig. 4.15). Hence, this sample is a good monitor of the functioning of the automatic frequency control (AFC) circuit of the spectrometer (Ludowise et al. 1991).

When the power incident on the resonator is very low, there is not enough reflected power to make the AFC circuit function properly. For example, Calvo et al. (2000) found that at temperatures below 4.2 K, incident powers of 0.2–1 nW were required in a study of the photosynthetic reaction center, and manual tuning was required to achieve a pure absorption EPR signal. The Q-band spectrometer used was locally built and used a TE_{011} brass cavity.

The latest Bruker spectrometers incorporate a DC AFC circuit to provide better stability at low incident powers and under some pulsed EPR conditions.

4.19 Resonator Design for Specific Samples

The amount of sample that can be put into a single-mode resonator is limited by the wavelength of the RF/microwaves. In the case of a non-lossy sample one can increase sample size with a minimal effect on Q. Alternatively, the Q can be increased substantially via special resonator design. However, Q will always be "sample-limited" in the case of a lossy (i.e., high dielectric constant) or conducting sample with finite dimensions. The resonator Q cannot be infinitely increased (e.g., by using superconducting resonators) because eventually increased source noise prevents the S/N from increasing further. Thus, for a non-lossy sample the ultimate S/N limit is the source noise. All of these considerations lead to the conclusion that the best S/N for a given sample will be achieved by designing a specific resonator for the sample. This is why there is such a wide array of resonators available.

4.20 Software

Modern spectrometers are software controlled, and contain multiple microprocessors to store calibration data and operate modules of the spectrometer under command from the main software program. The operator interacts with the spectrometer via the keyboard of a PC, with software that is vendor specific. Examples using the Bruker software to perform signal quantitation are provided in Appendix D.

4.21 Scaling Results for Quantitative Comparisons

Unless the spectra to be compared are obtained under exactly the same conditions, it is necessary to scale the area (double integral of the first-derivative signal) obtained for the unknown to compare it with the standard. For samples recorded at the same temperature and with the same spin number (e.g., $S = 1/2$) this involves:

1. Subtracting any background spectrum.
2. Correcting for modulation amplitude. Area scales linearly with modulation amplitude, even when spectra are overmodulated as long as the sweep width is wide enough to contain the entire EPR absorption envelope.
3. Correct for receiver gain. Area scales linearly with gain settings of the detector's amplifier. It is also important use a gain setting such that the signal plus noise come close to filling the full scale of the digitizer.
4. Correct for microwave power. Be sure to obtain spectra at power levels below saturation. Under these conditions, area scales as the square root of the incident microwave power.
5. Correct for Q differences. Area scales linearly with Q.
6. Correct for g-value differences. Area scales as g. Note that papers and books published prior to Aasa and Vanngard (1975) state incorrectly that area scales as g^2.
7. Correct for field scan width. For first-derivative spectra, the scan width correction factor is $(1/\text{sweep width})^2$.
8. When an integrating digitizer is used, the recorded signal area varies with scan time. This is often automatically normalized by the software in modern spectrometers.

If the software includes a "normalization" option, it is important to know which parameters are included in the normalization algorithm.

4.22 Signal Averaging

For a given time constant and A/D conversion time, the noise in a spectrum decreases, and hence the S/N increases, proportional to the square root of the number of scans averaged. The challenge for the operator is to choose the method of S/N improvement that is best for the sample and the environment. If the noise were white (i.e., truly random) all methods should give the same result. However, the real environment of a laboratory includes building vibrations (often of the sub-hertz to few hertz range). Sample and room temperature variations, power line variations and ventilation system air currents can all contribute to a slowly varying baseline that may distort the EPR spectrum and make double integration difficult. The weak pitch standard S/N test conditions are designed to reveal some of these problems. To eliminate these problems, it is usually better to perform signal averaging of multiple magnetic field scans at the fastest rate possible that still maintains line width integrity. Slow single scan experiments with a long filter time constant often reveal the low frequency baseline problems that were described above. The example in Fig. 4.16 shows spectra from the same sample recorded with the same total acquisition time except that the upper spectrum was acquired with several short scans and the lower spectrum used one long scan and long time constant. It is apparent that the lower spectrum has low frequency baseline deviations that could be mistaken for a background signal or part of the EPR sample.

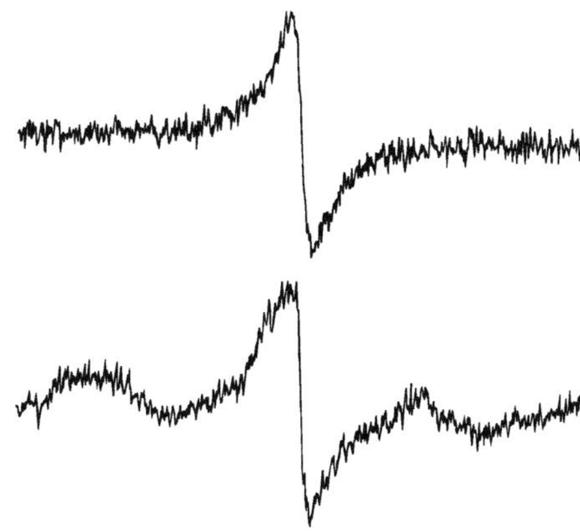

Fig. 4.16 The *upper spectrum* was acquired with short time constant and multiple-scan signal averaging. The *lower spectrum* was acquired with a long sweep time and long time constant

However, since this baseline deviation is random, it would not be possible to record a background spectrum for subtraction.

4.23 Cleanliness

Spins are everywhere. Most forms of "dirt" have an EPR signal. Many materials used for building construction contain easily measurable quantities of Mn(II). The cavity must be protected from "dirt." It is good practice to prohibit smoking in the spectrometer room, and to clean the outside of all sample tubes just prior to inserting them into the cavity. A tissue such as Kimwipe™ and a cleaner such as a residue-free electronic contact spray cleaner are useful for this purpose. Fingerprints will also transfer dirt to the cavity.

Inevitably, the cavity will become contaminated, possibly from a broken sample tube, causing significant background signals. Use the mildest cleaning that will remove the known or suspected contaminant. Often blowing out with clean air or compressed nitrogen gas is a good first step. Check the manufacturer's guidance on solvent compatibility. Some people use an ultrasonic cleaner, but this is not wise, as it may loosen critically torqued parts of the cavity. After cleaning, purge the cavity and waveguide with clean dry nitrogen gas to remove traces of solvent.

Chapter 5
Practical Advice About Crucial Parameters

The focus of Chap. 4 was the selection of operating parameters for recording EPR spectra. Chapter 5 will focus on how the interaction of the sample with the resonator and detection system impacts quantitation. Literature results on the achievable accuracy are discussed along with an explanation of the "tuning" procedure.

The general expression for CW EPR signal intensity is:

$$V_S = \chi'' \eta Q \sqrt{P Z_0} \tag{5.1}$$

where V_S is the signal voltage at the end of the transmission line connected to the resonator, η (dimensionless) is the filling factor (Poole 1967), Q is the loaded quality factor of the resonator, Z_0 is the characteristic impedance of the transmission line, and P is the microwave power to the resonator produced by the external microwave source (Rinard et al. 1999a, 2004). The magnetic susceptibility of the sample, χ'', is the imaginary component of the effective RF susceptibility. Optimizing the EPR measurement involves optimizing each of these crucial parameters. Key variables are χ'', Q and η. Each term in the expression for EPR signal voltage must be kept in mind during EPR measurements. B_1, Q, and η are discussed in more detail in Chaps. 6–8.

5.1 Crucial Parameters and How They Affect EPR Signal Intensity

Many spectrometer and sample parameters and interactions between the sample and the spectrometer affect the quantitative accuracy. Some of these parameters are controlled by the operator as discussed in Chap. 4. Others cannot be controlled, but must be known accurately to perform quantitative measurements. For example,

- The operator can control the preparation of the sample.
- The operator can select a sample and a resonator such that there is minimum perturbation of the microwave fields in the resonator. However, the resonator Q must be measured for accurate comparisons.

G.R. Eaton et al., *Quantitative EPR*,
DOI 10.1007/978-3-211-92948-3_5,

- The operator can make the reference standard such that its spectral width, intensity, and dielectric loss are similar to that of the unknown sample.
- However, the operator rarely can control the relaxation properties of the sample, but has to be sure that the sample is not power-saturated when making quantitative comparisons, or is saturated to the same extent as the reference standard.

The list of variables to consider is long, but the operator must always keep these in mind, and control or measure as many as possible.

- Microwave power and B_1 at the sample
- Modulation amplitude and frequency
- Receiver gain
- Scan time and detector time constant
- Magnetic field scan width
- Type of cavity
- Properties of any Dewar insert in the cavity (B_1 at the sample depends on the Dewar wall thickness)
- Physical size of sample
- Sample position in the cavity
- Dimensions and uniformity of the sample tube
- Dielectric properties of the solvent at the microwave frequency
- Detector current
- AFC offset
- Temperature of the sample and its effect on sample concentration, species dissociation, paramagnetism, and dielectric loss
- Microwave frequency (and magnetic field)

Some of these variables, and some EPR spectrometer operating aspects relevant to quantitative EPR are discussed in Wertz and Bolton (1972, Appendix D) and Weil et al. (1994b, Appendix E). Also valuable are the chapters: Bolton et al. (1972), Randolph (1972) and Alger (1968). In fact at the time of Alger's book, EPR was considered to be rather non-quantitative. A comparison study of standard samples was judged to exhibit errors of as much as 50% (Alger 1968, p. 205). Some of the factors affecting signal intensity were discussed by Bryson et al. (1975). In retrospect, it is likely that a large source of the uncertainty was the lack of measures of resonator Q. Dalal et al. (1981a) emphasized the importance of measuring Q, especially when using lossy solvents. Blakley et al. (2001) discussed the importance of matching the resonator Q for standard and sample. Fairly extensive reviews by (Eaton and Eaton 1980, 1990, 1997, 2005) provide additional references. As an example of the effect of lossy solvent on resonator Q, a 1 mm diameter capillary tube containing water reduced the Q of a rectangular cavity from ca. 3000 to 500–600. Since the EPR signal is proportional to resonator Q, an attempt to make a quantitative measurement of a 1 mm aqueous sample relative to a non-lossy standard could introduce an error of a factor of 5 or so. However, an accurate comparison can be regained if one scales the amplitude results by the resonator Q-factor difference for the two samples.

It is important to consider the age of the EPR spectrometer. In the early days of EPR there was little focus on quantitative accuracy of spin concentrations. Goldberg (1978) documented the uncertainties in settings on older Varian spectrometers and their effect on signal levels. Better results were obtained by Yordanov and Ivanova (1994b). The rather discouraging view of intra- and inter-laboratory comparisons presented in these older papers have been relieved substantially by the attention paid to resonator Q, sample positioning, reference standards, and dramatic improvements in EPR spectrometers.

5.2 What Accuracy Is Achievable?

With modern spectrometers, and careful attention to experimental conditions, inter-laboratory comparisons have achieved 3% standard deviation on samples with internal standards (Gancheva et al. 2008). Alanine dosimeters containing Mn^{2+}/MgO as an internal standard were measured on 12 instruments of 6 different models by three manufacturers, and run by 10 operators. This 3% standard deviation was obtained only after a preliminary comparison showed the need to use a procedure that involved low power, low modulation amplitude, and a combination of sweep time and time constant that provide a distortion-free spectrum. In another study Yordanov et al. (1999) reported that the same operator on the same instrument can achieve about $\pm 2\%$ reproducibility on the same sample. However, if the measurement was made relative to an internal standard, the standard deviation was reduced to 0.5%.

Even with modern instruments, some caution needs to be observed. For example, incident microwave power is controlled by a rotary vane attenuator. In older instruments the rotary vane attenuator was set manually by rotating a knob which caused the gears in the device to turn. In modern computer controlled X-band spectrometers the same type of rotary vane attenuator is used, but the gears are turned by a motor (instead of by a hand). Ultimately, the accuracy of the attenuation is dependent on the small backlash of the gears in the device. For the most accurate quantitative work it is important to approach the final attenuation setting from the same direction in all of the experiments. Microwave sources do have finite lifetimes, and the power output could change with time. So, periodic calibration either with a power meter or with a standard sample is a good practice.

It has often been said that the system should be allowed to "warm up" before recording spectra. Although this advice is a holdover from the vacuum tube era, it actually retains some validity for different reasons. The sensitivity of a crystal detector is temperature dependent, so for the most accurate quantitative results, the crystal should be at the same temperature in compared measurements. In fact, the Bruker spectrometers specifically designed for quantitative EPR have provision for temperature control of the detector. Microwave detectors can fail quickly due to high power burnout, but more commonly, the sensitivity slowly degrades over time due to occasional, non-lethal, excess incident power. Once again, periodic calibration with a standard sample is useful to monitor this aspect of spectrometer performance. Detector diode response deviations are small and fairly uncommon on modern instruments, but cannot be ignored for the most accurate work.

Quantitative EPR depends, ultimately, on the analytical balance and on volumetric techniques such as calibrating glassware, some of which is an art disappearing from standard college curricula. Unless precision quartz sample tubes are purchased, for accurate work the sample tubes need to be calibrated just like any other piece of analytical glassware. For example, one can weigh a measured length of solvent in the tube at a known temperature to measure the internal volume of the tube. There are uncertainties due to the rounded bottom of the tube and meniscus at the top of the column of fluid, however, these approximately compensate for one another. Thus, weighing the empty tube, then adding liquid and weighing again, yields the mass of a measured length of fluid. Tables of density vs. temperature in the Handbook of Chemistry and Physics can be used to derive the volume of the tube. It is a bit more difficult to reproducibly put powdered or microcrystalline material in a tube. The main concern is reproducibility, so each operator has to develop a procedure. For example, add a couple of millimeters of loose powder at a time, followed by tapping in a routine way on a hard rubber stopper; then add more powder and repeat until the column of material is the desired length. The packing density is measured by weighing the tube before and after filling and is strongly dependent on the method of packing the powder.

5.3 A More In-Depth Look at Adjusting the Coupling to the Resonator in the "Tuning" Procedure

Adjusting the coupling to the resonator can be one of the most time-consuming aspects of preparing to obtain an EPR spectrum. To the novice operator it may seem like a mysterious ritual, so this section attempts to remove some of the mystery. The goal stated in most operating instructions is to have the resonator "critically coupled" to the transmission line. This means that the iris coupling assembly of a cavity (Fig. 1.9), or the antenna coupling assembly of a dielectric or split ring resonator, will perform as a transformer that transforms the impedance of the resonator to the impedance of the transmission line (waveguide or coaxial line). At "critical coupling" (i.e., when the impedance is properly matched between the resonator and transmission line) the microwave power incident on the resonator is completely absorbed in the resonator, and none of the power is reflected to the detector. The tuning "dip" is a plot of reflected power vs. microwave frequency. Ideally, zero power is reflected at critical coupling, so the dip should be as deep as it will go.

At EPR resonance, the impedance of the resonator changes and some of the incident power is reflected to the detector in the bridge. This power becomes the EPR signal. This simple picture is complicated by the fact that the EPR bridge is based on an imperfect circulator. Although most of the microwave power follows the circular route from source to resonator to detector (ports 1, 2, and 3 in Fig. 1.6), about 1% of the power incident on port 1 goes directly to port 3 (detector) instead of to port 2 (resonator). Thus, the power reaching the detector is the sum of the ideal EPR signal model for a critically coupled resonator and the microwave power that goes through this direct leakage path (i.e., directly through port 3 from port 1). Since both signals are affected by the same attenuator, the leakage power and the power that is incident on the resonator increase together. The combined power from the reference arm, the leakage and that reflected from the cavity provide the bias power to allow the detector to perform in its linear response region. The typical detectors used in EPR spectrometers, require a current of about 200 μA. The spectrometer functions best for quantitative EPR when there is only a very small change in detector current as the spectrometer goes through EPR resonance. If a change in the detector current is observed at resonance, the EPR signal is much too strong for quantitative results.

From an engineering perspective, by turning the iris (Fig. 1.9), the operator is actually adjusting the resonator, circulator, and transmission line to the critical coupling condition as a complete unit. The resonator, itself, may be slightly away from critical coupling. Slight mismatch of the resonator at some position of the coupling device causes the reflected power to cancel the leakage through the circulator, with the result that the detector current will stay almost constant as the power incident on the resonator is changed by changing the attenuation – i.e. the goal of the "coupling procedure".

To properly bias the crystal detector some of the source power is routed around the circulator in the reference arm. The reference arm microwave phase controller adjusts the phase of the reference arm power to be coherent with the phase of the power being reflected from the cavity during the EPR absorption. In the tuning procedure the final phase adjustment seeks to maximize the detector output, and hence maximize the EPR signal.

Chapter 6
A Deeper Look at B_1 and Modulation Field Distribution in a Resonator

The EPR signal is proportional to the microwave B_1 at the sample, which is proportional to $\sqrt{P}$. Consequently, it is important to carefully examine the distribution of B_1 over a sample of finite size, such as a standard liquid or powdered sample in a 4 mm o.d. quartz sample tube. In the typical EPR experiment that uses magnetic field modulation and phase-sensitive detection, the integrated signal intensity is proportional to the modulation amplitude at the sample. Therefore, it is also important to consider the distribution of modulation amplitude over the sample. The details of these two factors are discussed in this chapter. This chapter also includes discussion of sample size, issues related to automatic frequency control (AFC) for very narrow signals, and cell geometries for aqueous samples.

6.1 Separation of B_1 and E_1

It is the microwave magnetic field (B_1) that induces the EPR transitions that are detected in EPR spectroscopy. Also associated with B_1 is the microwave electric field (E_1). The E_1 can induce rotational transitions in the sample, thereby generating heat. This phenomenon should be familiar to readers from the effects of a microwave oven on food. This microwave absorption contributes to additional energy dissipation and thereby reduces the resonator Q (see Chap. 7). To avoid excessive interaction of the sample with the E_1 field (and resultant Q lowering), it is important to position the sample in a region of the cavity with high B_1 and low E_1.

For cavities, there is a natural separation between B_1 and E_1 because upon resonance, a standing wave is excited within the cavity. Standing electromagnetic waves have their electric and magnetic field components exactly out of phase, i.e. where the magnetic field is maximum, the electric field is minimum and vice versa. The spatial distribution of the electric and magnetic field amplitudes in the commonly-used TE_{102} rectangular mode cavity is shown in Fig. 6.1.

DOI 10.1007/978-3-211-92948-3_6,

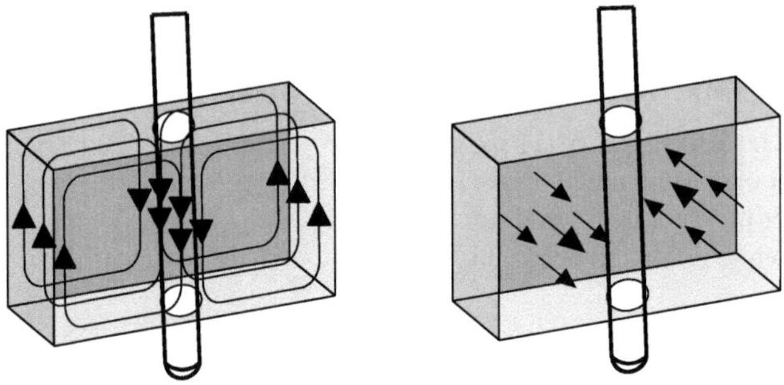

Fig. 6.1 Placement of the EPR sample in a TE_{102} rectangular cavity for the maximum B_1 and minimum E_1

The spatial separation of the electric and magnetic fields in a cavity is used to great advantage. When the sample is placed in the electric field minimum and the magnetic field maximum, the biggest signals and the highest Q are obtained. Dielectric properties of the sample can also change the field distribution. Cavities are specifically designed to provide optimal placement of the sample with regard to B_1.

6.2 Inhomogeneity of B_1 and Modulation Amplitude

The microwave B_1 field is inherently nonuniform in a microwave cavity because the resonant condition necessitates standing waves, which must have minima and maxima. Modulation amplitude also is not homogeneous in the resonator. There have been many studies of the B_1 and modulation distributions in resonators. Early studies that informed operators of the major considerations were performed on the Varian E231 cavity. The early literature is summarized in Eaton and Eaton (1980) and experimental guidance is given in Eaton and Eaton (2005). Dalal et al. (1981a) describes the effects of lossy solvents on B_1 and modulation amplitude distribution. More et al. (1984) describe the simulation of power saturation curves of line samples considering B_1 and modulation distributions. Additional information is given in (Schreurs et al. 1960; Kooser et al. 1969; Mailer et al. 1977). More recent studies have examined the Bruker TE_{102} cavity resonator (Mazúr et al. 2004, 2005; Mazúr 2006; Yordanov and Slavov 1996; Yordanov and Christova 1997; Yordanov and Genova 1997; Yordanov et al. 1999 and references cited therein).

The nonuniformity in B_1 and modulation amplitude have resulted in many papers describing mechanical devices intended to precisely and reproducibly position samples in the resonator (see, for example, Chang 1974). The sample also affects the microwave field (Rages and Sawyer 1973; Dalal et al. 1981a).

Figure 6.2 provides an example of a device to monitor the dependence of signal intensity on sample position in the resonator. Such an experiment provides a

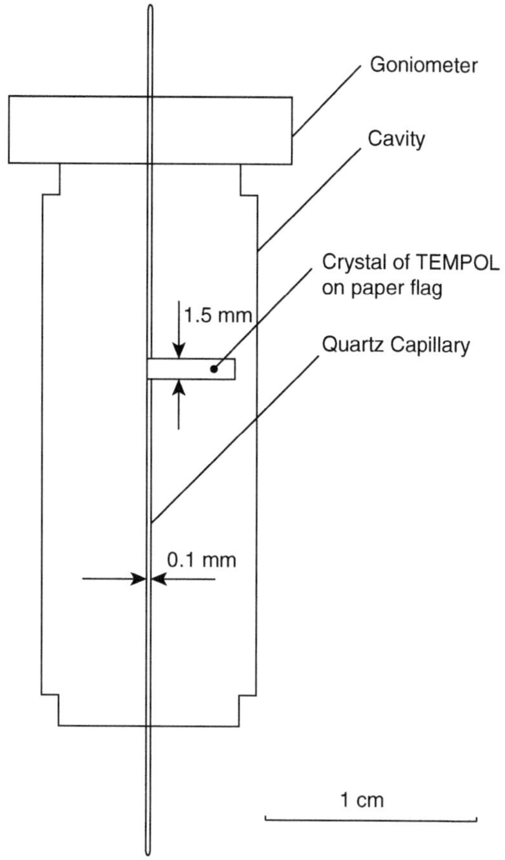

Fig. 6.2 Device to position a sample for measuring signal amplitude as a function of position in the cavity. From Nagy (1994)

measure of the product of B_1 and magnetic field modulation amplitude. Mazur and coworkers and Yordanov and coworkers characterized the Bruker TE_{102} rectangular "ST" resonator and the TE_{104} double resonator. The effect of B_1 distribution and modulation amplitude on quantitative EPR was measured for 14 commercial cavities (Yordanov et al. 2002). Due to differences in construction, the B_1 and modulation field distributions along the length of a sample could be similar, or one could be longer than the other. Table 6.1 lists a major topic for each of the papers in this series. The papers contain more detail than is indicated in the table.

The effect of distribution of B_1 and modulation on the EPR signal in a Bruker ER4119HS resonator is shown in Fig. 6.3.

The nonuniformity of the B_1 field and modulation amplitude also affect the measurement of microwave power saturation (Mailer et al. 1977). Comparison of experiments between labs will not agree unless corrections are made for the variations in B_1 and modulation amplitude over the sample region of the cavity. Different types of resonators have different distributions of B_1 and modulation field over the sample.

Table 6.1 Characterization of TE_{102} and TE_{104} cavity resonators

Topic	Reference
Effect of shape of sample and sample position	Nagy and Placek (1992)
Alignment procedure for sample in TE_{104} cavity	Mazúr et al. (1996a)
Movement of line samples into TE_{104} cavity	Mazúr et al. (1996b)
TE_{104} rectangular double resonator; differences in sample shape	Mazúr et al. (1997c)
TE_{104} and TE_{102}; line sample moved along axis	Mazúr et al. (1997a)
Alignment procedure for sample in TE_{104} cavity	Mazúr et al. (1997b)
Line sample in TE_{102} or TE_{104} cavity	Mazúr et al. (1998)
Radial and angular dependence in TE_{102} cavity	Mazúr and Valko (1999)
Radial and longitudinal dependence in TE_{102} and TE_{104}	Mazúr et al. (2000)
Large cylindrical samples with variable internal diameter and lengths of sample in TE_{102} and TE_{104}	Mazúr et al. 2001)
A dozen useful tips – sample positioning etc.	Mazúr (2006)
Effect of quartz insert in TE_{102}	Yordanov and Slavov (1996)
TE_{102} and TM_{110}; length and diameter of samples	Yordanov and Genova (1997)
Distribution of modulation and microwave B_1	Yordanov et al. (2002)

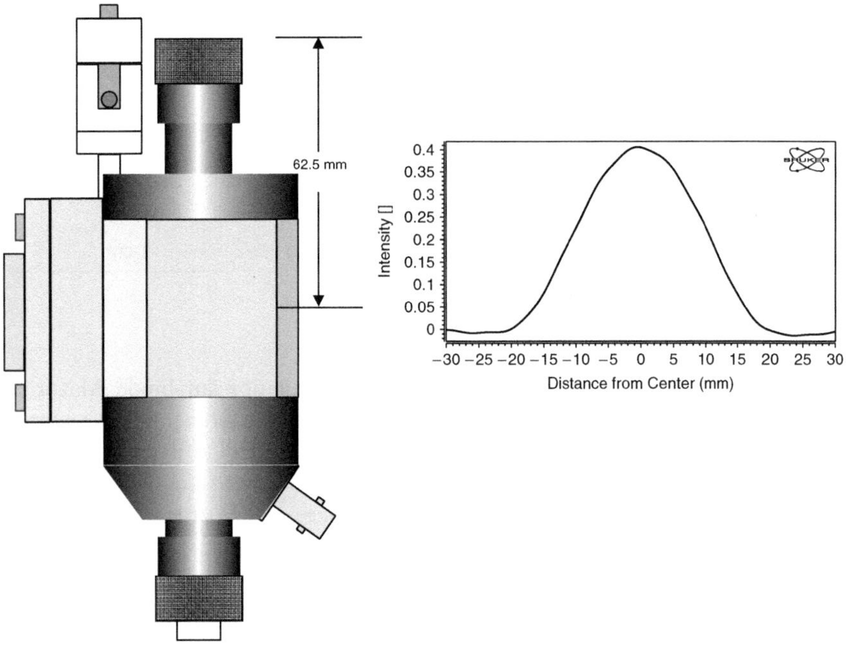

Fig. 6.3 Signal intensity distribution for a Bruker ER4119HS resonator

Recently, the Hyde lab has described "uniform field" resonators, in which there is a sample region where the B_1 field is uniform over a larger fraction of the resonator than in standard cavity or loop gap resonators. These are designed especially for aqueous samples (Mett et al. 2001; Anderson et al. 2002; Hyde and Mett 2002; Hyde et al. 2002; Mett and Hyde 2003; Sidabras et al. 2005).

The B_1 distribution in the resonator affects the allowed size, shape, and placement of the sample in the resonator. Ideally, one wants a uniform B_1 field over the sample. In addition, if the sample is lossy, one wants negligible E_1 field in the sample. In a TE_{102} cavity resonator there is a nodal plane at which B_1 is maximum and E_1 is minimum (Fig. 6.1). This is the normal sample location. In cylindrical resonators, the node is the central axis of the cylinder. Split ring, loop gap, and other such lumped circuit resonators approximately separate the B_1 and E_1 fields, but have a capacitive gap that results in a fringing E_1 field near the sample. Various bridged loop gap resonators are designed to minimize the fringe E_1 field. This E_1 field is why frequency and Q are so strongly dependent on sample placement in these type resonators.

6.3 Sample Size

A sample that is too small can be difficult to quantitate because unless it is highly concentrated, it will likely have low S/N. The low S/N also results in resonator background signals and baseline slopes that complicate analysis of the EPR spectrum. In some cases a sample can also be too large, as reported by Vigouroux et al. (1973) and by Goldberg and Crowe (1977). Vigouroux et al. (1973) showed line broadening when a TE_{011} cylindrical cavity with $Q = 1{,}500$ contained more than about 10^{19} spins of DPPH and more than about 3×10^{20} spins of Mn(II). Proportionately more spins are acceptable for samples with hyperfine splitting such as Mn(II). Goldberg and Crowe (1977) made more precise measurements of widths and amplitudes and found nonlinearities on the order of 1% when samples of ca. 3.4×10^{18} spins of $MnSO_4{\cdot}H_2O$ or ca. $3.6 \times 5\ 10^{17}$ spins of DPPH were used in a TE_{102} cavity with a loaded Q of about 3,000. Although not tested by Goldberg and Crowe, proportionately fewer spins will cause problems with extremely narrow-line samples, such as lithium phthalocyanine (Smirnov et al. 1994; Ilangovan et al. 2004) and triarylmethyl radicals (Ardenkjaer-Larsen et al. 1998).

If an unknown and a known sample are the same size and are long ($>$ 2 cm for X-band), the precise position becomes less of an issue for quantitating the unknown.

6.4 AFC Considerations

For concentrated paramagnetic substances with many spins per gram, (e.g., solid DPPH, BDPA, LiPc, etc.) a barely visible sample can easily overload the spectrometer. In extreme cases the operator can see changes in the detector current and AFC offset voltage as the field is swept through resonance. Any noticeable effect on the AFC or detector current is much too large for proper quantitative EPR spectroscopy. Quantitative EPR assumes that the microwave absorption in the sample is a small perturbation on the resonator Q (see Chap. 7). Microwave absorption by the sample is a change in the resistance term in the denominator of the equation that defines the

Q-factor. This change in resistance is what causes the EPR signal, but it has to be very small to be quantifiable. For a sample with an extremely narrow line, attempting to measure a sample that has too many spins per gram can cause a disturbance in the AFC that creates an even narrower line. Figure 6.4 is an example. Turning off the AFC requires very careful phasing, but gives a more accurate line shape.

Operating with a field/frequency lock often is the preferred way to stabilize magnetic field/frequency for recording narrow lines. However, if the signal is strong and very narrow, some rather bizarre spectra can be recorded, such as shown in Fig. 6.5, which was obtained with the same trityl sample as were the spectra in Fig. 6.4. The hypothesis is that the intense narrow signal causes the F/F lock to oscillate when searching for a proper match. When using an external resonator to lock the microwave source, the shape of the spectrum depends strongly on the exact frequency of the external stabilizer resonator.

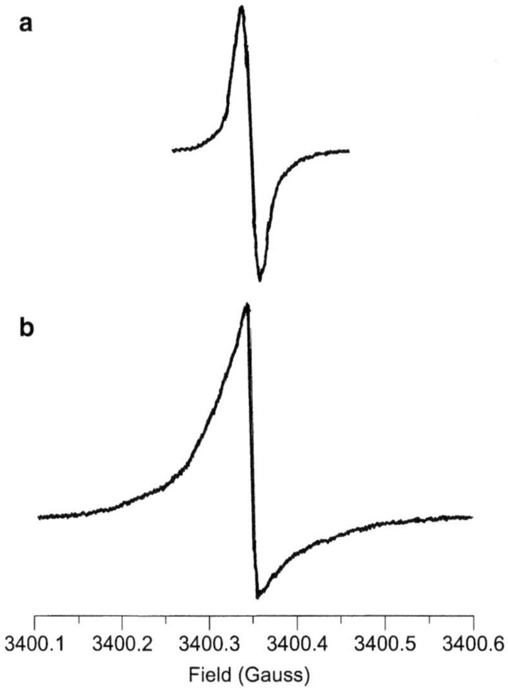

Fig. 6.4 The two spectra in this figure are of the same sample of 0.2 mM trityl (Nycomed deuterated symmetric trityl) in a Teflon tube inside a 4 mm quartz tube, after removal of O_2 by passing gaseous N_2 over the Teflon tube. The spectrum on the top matches expectations for this species, which has a line width of ca. 26 mG and ^{13}C hyperfine lines close to the center line. This shape could be obtained without AFC lock, or with the AFC locked to an external cavity (the external stabilizer cavity used for pulsed EPR in an E580 bridge was used in this case). 10 mG 7 kHz magnetic field modulation was used, and the microwave power was attenuated 56 dB from 200 mW (0.5 μW). Note that the total scan width is 200 mG. The spectrum on the bottom was obtained with normal AFC lock and 500 mG scan width, and shows a distorted lineshape. The F/F lock was not used for either spectrum

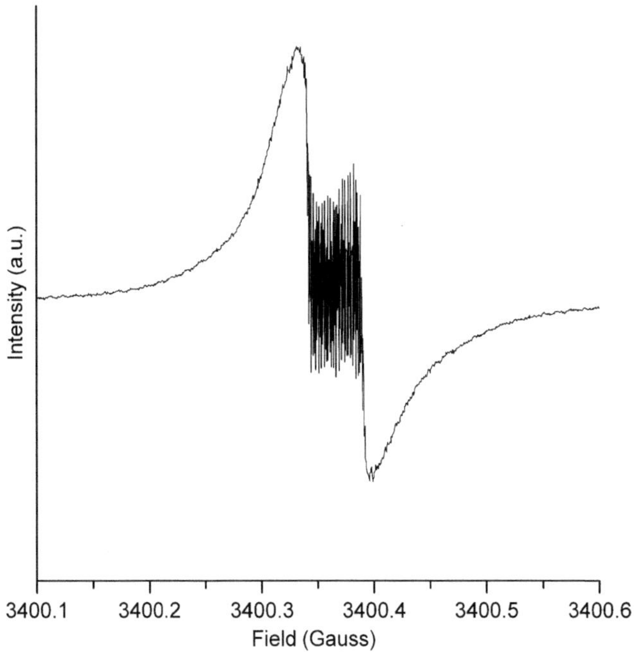

Fig. 6.5 This strange X-band spectrum was obtained while attempting to use the field/frequency lock to record a narrow-line 0.2 mM deuterated symmetric trityl spectrum. The rapid oscillations are attributed to the effect of the very strong narrow signal on the ability of the F/F lock to find the proper match

These phenomena have been observed with many very-narrow-line EPR spectra. Great effort is required to avoid power saturation, modulation broadening, AFC effects, field–frequency lock distortions, and phase distortions for very-narrow-line signals. Spectra similar to that in Fig. 6.4 have been published and attributed to spin systems with very narrow lines. The problem is inherent in AFC systems, including the new Bruker DC-AFC systems.

6.5 Flat Cells

Since improved S/N is a constant goal in EPR, the operator wants to put as large an aqueous sample in the resonator as possible. Decades ago it was shown that large aqueous sample volumes could be measured using a flat-shaped quartz cell with dimensions determined empirically. It was found that the Bruker ER 4103TM (TM_{110}) cavity with a large, specially designed flat cell gives a sensitivity about a factor of ~4 higher than an ER 4102ST (TE_{102}) cavity using a ~1 mM i.d. capillary tube. With this optimized flat cell and the appropriate positioning devices, the largest lossy samples can be studied with this cavity. Precise positioning of flat cells is facilitated by specially designed holders that attach to the bottom and top of the cavity. Using the modern Bruker ER 4119HS cavity, an almost identical enhancement in sensitivity is achieved using a 10 mm flat cell; which is less orientation dependent. It was observed empirically by Hyde (1972) and by Eaton and Eaton (1977) that rotating the flat cell 90° – perpendicular to the nodal plane – also yielded high quality results. The perpendicular orientation of the flat cell was originally observed in a Varian cavity, but has often been found to be the preferable orientation for a 10 mm flat cell in the Bruker ER 4102 ST cavity. In addition, sandwiches of flat cells worked better than single flat cells. Dividing the aqueous sample into smaller portions minimizes Q spoiling and provides better S/N than one large water sample. This principle also forms the basis for the Bruker AquaX™ sample cell which consists of many small diameter cylindrical tubes. This phenomenon has been simulated using finite element methods by Hyde and coworkers (Mett et al. 2001; Anderson et al. 2002; Hyde et al. 2002; Mett and Hyde 2003; Sidabras et al. 2005).

The expense (and fragility) of commercial quartz flat cells cause many investigators to use capillary tubes for routine studies of aqueous samples at X-band. Rectangular glass capillary tubing that can be used for disposable flat cells is now available from VitroCom (New Jersey) as "large ratio rectangle tubing" in borosilicate or clear fused quartz. The 0.4 by 8 mm inner dimension size is convenient.

6.6 Double-Cavity Simultaneous Reference and Unknown

A double-cavity resonator, made by joining two TE_{102} cavities together is another idea for providing accurate comparisons between a reference standard and unknown samples. Since the frequency will be the same, and the magnetic field is sufficiently homogeneous over the larger double cavity, the resonance field also is the same for both sample and reference. These features make a double-cavity useful for measuring relative g-values when it is more convenient to have separate reference and unknown samples rather than using concentric or internal standards. However, the perturbation of the microwave field by the sample can result in different magnitude of B_1 at the standard and the unknown. This effect is sketched in Fig. 6.6. A quartz Dewar was inserted into one sample port which caused the B_1 field in the other sample port to shift by 1–2 mm. Mazúr et al. (2003) showed that having a Dewar placed in one cavity could decrease the signal from a sample in the other cavity.

6.7 Summary

The *absolute* determination of the number of spins in a sample has been described as one of the most difficult EPR measurements. However, the same is true of quantitation in visible spectroscopy. The extinction coefficient reported for DPPH

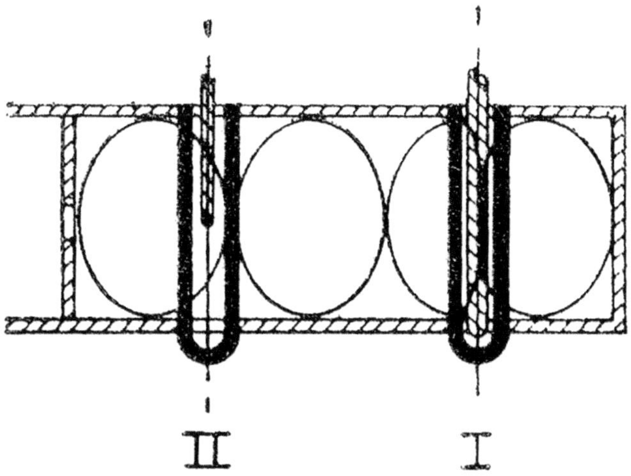

Fig. 6.6 Microwave magnetic field compression and shift of the maximum field in a dual cavity with two samples. From Casteleijn et al. (1968)

varies by as much as 10%. Very careful end-to-end measurements have to be made with regard to the gain of the spectrometer, and the resonator has to be very carefully characterized (Rinard et al. 2002a, b, 2004). There was good agreement between calculated and measured signal intensities reported in the abovementioned papers as a result of an enormous number of measurements characterizing every feature of the spectrometer. Note, for example, that it is difficult to measure gains and losses of many microwave components to better than 1 dB without a very carefully calibrated test bench, but 1 dB error in power ratio has 12% effect on the EPR signal (see Appendix G for further discussion). Consequently, one compares an "unknown" sample with a "known" sample (see Chap. 11). The latter should have been prepared using the quantitative methods of an analytical chemistry laboratory. Further, to compare the signals from two samples it is essential to know the filling factor, η, the Q, and the power incident on the sample to within the accuracy of the desired result. Because B_1 and the modulation amplitude vary across the resonator, reproducible positioning of the sample in exactly the same size tube is essential for the highest accuracy. Unless the sample is in the same solvent, in the same size tube, in the same position in the resonator, the Q is not likely to be the same, and would have to be measured by one of the methods discussed in Chap. 7. The comparison also assumes that both samples are in the linear response region with regard to microwave power. That is, the signal response is strictly linear with B_1 and does not exhibit any microwave power saturation. So, it is also common practice to characterize the power saturation behavior for both samples before proceeding with quantitative comparisons. The microwave frequency must also be the same. Many of these conditions are difficult to achieve with a high-filling-factor resonator, because in such a resonator the sample has a large effect on the frequency and Q. Therefore, most quantitative EPR studies have been done in cavity resonators, and the most carefully characterized is the standard TE_{102} rectangular resonator.

There are cases where relative EPR amplitude measurements are used in industrial QA/QC applications with very stringent demands placed on quantitative results. If there are many samples that are the same size, physical properties, etc., it is possible to get very accurate relative EPR amplitude measurements. With the use of built-in amplitude reference standards, overall measurement uncertainties of 3% have been achieved. The Bruker BioSpin e-scan benchtop EPR units have been tailored for specific applications often requiring relative quantitative measurements with high sample-to-sample precision, accuracy and throughput. The development of these systems followed the principles outlined in this chapter.

Chapter 7
Resonator Q

The quality factor, Q, of a resonator is of vital importance for quantitative EPR spectroscopy. Many problems with the reproducibility of EPR intensities within a lab, or among labs, are due to different values of the resonator Q. To perform quantitative comparisons of EPR signals, it is necessary to either measure Q, or to perform the comparison in such a way that Q is kept constant (e.g., via same sample size, dielectric loss properties, coupling of the resonator, etc.).

Equation (7.1) shows that the signal voltage is directly proportional to Q.

$$V_S = \chi'' \eta Q \sqrt{PZ_0} \tag{7.1}$$

Resonance means that the resonator stores the microwave energy. Therefore, at a resonator's "resonant frequency," no microwaves are reflected if it is "critically coupled." In reality only a small amount of microwave energy is stored in the resonator. Most of the incident microwave power is dissipated in the resistance of the resonator. Resonators are characterized by their quality factor (or Q) which indicates how efficiently the cavity stores microwave energy. The Q, is defined as:

$$Q = \frac{2\pi \times (\text{ energy stored })}{\text{energy dissipated per cycle}} = \frac{2\pi(\text{frequency})(\text{energy stored})}{\text{energy dissipated}}, \tag{7.2}$$

where the energy dissipated per cycle is equal to the amount of energy lost during one microwave period. Energy is lost because the microwaves generate electrical currents in the resonator which in turn generate heat. As Q increases, the sensitivity of the spectrometer increases. Q factors can be measured easily because there is another way of expressing Q:

$$Q = \frac{\nu_{res}}{\Delta \nu}, \tag{7.3}$$

where ν_{res} is the resonant frequency of the cavity and $\Delta\nu$ is the width of the cavity dip at half of its height for reflected power (Fig. 7.1). Broad dips correspond to low Q factors and narrow dips correspond to high Q factors.

G.R. Eaton et al., *Quantitative EPR*,
DOI 10.1007/978-3-211-92948-3_7,

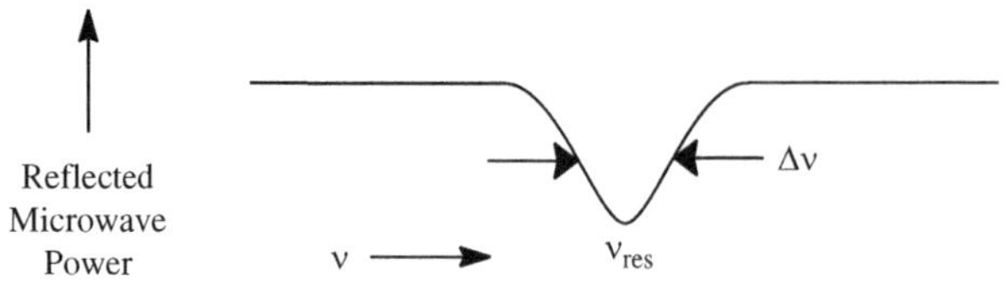

Fig. 7.1 Reflected microwave power from an EPR resonator

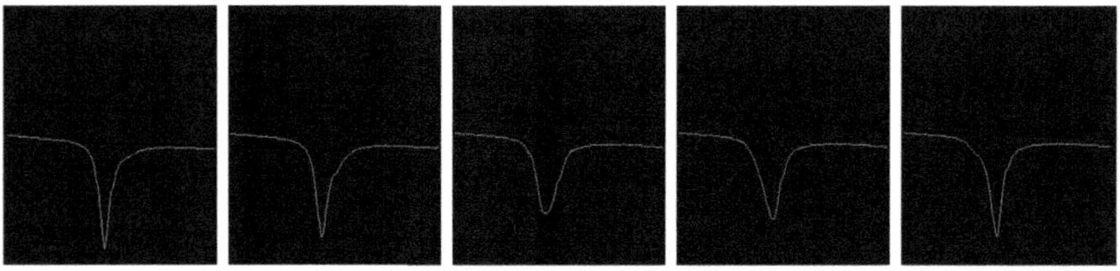

Fig. 7.2 Changes in the resonator "dip" due to inserting and removing a lossy sample

EPR spectrometers have a display mode called "Tune" in which the microwave frequency is swept on the horizontal axis and the reflected microwaves are detected to form the data for the vertical axis. The dip (or resonant frequency) of the resonator can be found from this display. When a "lossy" sample is inserted into a resonator, the Q decreases and this is easily seen from the broadening of the resonator dip (Fig. 7.2). If the sample loss is too high, it will not be possible to critically couple the resonator, and a smaller sample is required.

The effect on resonator Q from inserting a lossy sample is shown in the series of snapshots of the tuning mode of the spectrometer (Fig. 7.2).

7.1 Conversion Efficiency, C′

It is the B_1 that drives the EPR transition. The EPR intensity grows linearly with B_1 if the spectrum is not saturated (see Fig. 4.5). What the operator sets is the microwave power. How are B_1 and power related? B_1 increases proportional to the square root of power.

$$B_1[\mathrm{G}] = \frac{C'\sqrt{P[\mathrm{W}]}}{\sqrt{\Delta\nu[\mathrm{MHz}]}}, \tag{7.4}$$

where P is the microwave power in watts. Δν is the resonator bandwidth in MHz that has already been mentioned in the discussion of Q. C′ is the conversion efficiency.

The scaling between B_1 and P (7.4) is resonator dependent. Although decreasing the bandwidth (i.e., increasing the Q) improves conversion efficiency, this may have undesirable consequences for certain types of experiments and Q may be affected by the dielectric properties of the sample. The conversion efficiency is a characteristic of a resonator and a crucial parameter that depends on many complicated factors. Optimization of C' is one of the most difficult challenges in designing resonators.

As a consequence of an electrical engineering principle called "reciprocity" the EPR signal detected for a given B_1 is proportional to the conversion efficiency (C'). Since B_1 is also proportional to C', at a given incident microwave power, the EPR signal is proportional to $[C']^2$. As a general rule, C' increases as the resonator size decreases. In essence, decreasing resonator size concentrates and focuses more of the microwave power on the sample. For a loop-gap resonator, B_1 is proportional to $1/\sqrt{(\text{length} \times \text{diameter})}$.

It is vital for quantitative EPR comparisons that B_1 and not the microwave power is the parameter used for comparing results for two samples. Just because the same power was used in the same resonator does not mean that B_1 was identical for both measurements: the effect of the sample's dielectric properties on the B_1 distribution and on the Q could be different.

7.2 Loaded Q and Unloaded Q

For EPR the most useful Q-factor expression is called the "loaded Q" of a resonator. This term is sometimes used incorrectly in the magnetic resonance literature to mean the Q of a resonator "loaded" with a lossy sample, such as an animal. The electrical engineering definition of "loaded Q" is the Q of a resonator coupled to a transmission line. The "unloaded Q" is the term used for the resonator when it is not coupled to a transmission line and is twice the value of the loaded Q. Sometimes subscripts L and U are used to denote the loaded and unloaded Q values, but since the loaded Q is so commonly used, the L subscript is often omitted. Caution should be exercised in citing Q values unless they are clearly defined.

There are many ways to express the Q of a resonator. Each provides useful insight, so several expressions are provided here.

In terms of an RLC circuit, the resonator Q is given by:

$$Q = \frac{\omega_0 L}{R}. \tag{7.5}$$

This is the unloaded Q. The loaded Q is half of this value when the resonator is critically coupled. Rewriting this equation with the denominator equal to the sum of the resistance of the resonator and of the sample gives:

$$Q = \frac{\omega_0 L}{R + r_{sample}}. \tag{7.6}$$

This equation nicely shows that the lower the resistance of the resonator (R) and of the materials in it, r_{sample} (e.g., less lossy sample), the higher the Q.

The inductance of a cylindrical resonator is related to the length, z, and the cross sectional area, $A_R = \pi d^2/4$ (d is the diameter of the loop)

$$L = \mu_0 \frac{A_R}{z}. \tag{7.7}$$

Skin depth is

$$\delta = \frac{1}{\sqrt{\mu_o \pi \nu \sigma}}, \tag{7.8}$$

where σ is the conductivity of the material of the resonator and $\mu_0 = 4\pi \times 10^{-7}$ $T^2J^{-1}m^3$ is the permeability of vacuum. So, the resistance, including skin effect, of the resonator loop is

$$R = \frac{\pi d}{z}\sqrt{\omega \frac{\mu_0}{2\sigma}}. \tag{7.9}$$

The angular frequency of the resonator is given by

$$\omega = 2\pi\nu = \frac{1}{\sqrt{LC}}. \tag{7.10}$$

In loop gap resonators the L and C can be visualized geometrically and the capacitance can be calculated, neglecting fringing, using:

$$C = \frac{A\varepsilon}{d}, \tag{7.11}$$

A is the area of the plate of the capacitor, d is the distance between plates, ε is the permittivity of the material between the plates of the capacitor. Permittivity is the dielectric constant x ε_0 ($\varepsilon_0 = 8.854 \times 10^{-12}$ F m^{-1})

In EPR resonators with a capacitive gap, the dielectric in the gap is usually air (or N_2 or He gas during low temperature operation), but it is common to use Teflon in the gap to increase the capacitance and thus lower the frequency of a lumped circuit resonator. The main problem with using a dielectric such as Teflon in the capacitive gap is that its dielectric constant is very temperature dependent.

In actual resonator design, modifications of the above formulae are used that account for edge effects of finite length inductors and finite area capacitors, but the formulae here give a good approximation. It is much harder to estimate the capacitance of narrow gaps and specially shaped gaps such as in the Bruker split ring resonator.

Thus, knowing the properties of the materials of the resonator and the size, one can calculate Q (Poole, 1967). However, it is easier and more accurate to measure Q than to calculate it from first principles.

7.3 Relation of Q to the EPR Signal

From a classical viewpoint, the EPR signal is a voltage that is generated by spins that are in the *xy* plane of the resonator (the static magnetic field is in the *z* direction). For CW EPR, this is often viewed as absorption of microwave energy at resonance increasing the effective resistance in the denominator of the equation for Q (7.6). The change in resonator Q results in a change in the microwave power reflected from the resonator to the detection system, because the resonator is no longer critically coupled. This description reinforces the concept of equivalent resistance in the denominator of the equation for Q. This change in reflected power is what is converted into the "EPR spectrum" and interpreted by the spectroscopist. Similarly, absorption of microwaves by a "lossy" sample also causes an increase in the resistance term for the denominator of the equation defining Q, which lowers the Q, and hence, lowers the EPR signal intensity.

Alternatively, the origin of the EPR signal can be described in terms of precessing spins. The microwave power absorbed by the sample creates a net magnetic vector rotating at the Larmor frequency. The magnetic field of the precessing spins produces a voltage in the resonator, and the resultant voltage couples into the transmission line reciprocally to the way the microwaves got to the resonator. Since the fields due to the electron spins are encoded by the field/frequency resonance, the resultant microwave signal at the detector is the EPR signal. Sometimes one speaks in terms of the currents induced in the conductors of the resonator by the spin magnetic field, but the more general discussion in terms of the fields themselves works for both metallic and dielectric resonators.

7.4 Contributions to Q

There are many contributions to the measured loaded Q (Q_L) of an EPR resonator. They add as reciprocals (Dalal et al. 1981a).

$$\frac{1}{Q_L} = \frac{1}{Q_U} + \frac{1}{Q_\varepsilon} + \frac{1}{Q_r} + \frac{1}{Q_\chi} + \frac{1}{Q_\mu}. \tag{7.12}$$

The contributions to Q include:

- Q_U is the unloaded Q, with loss attributable only to the resistance of the walls of the resonator.
- Q_ε is the effect of dielectric losses on Q.
- Q_r is the effect of power lost through the cavity coupling holes or due to conducting walls less than many skin depths. (Q_r is also called the radiation quality factor.)
- Q_χ is the effect of power absorbed by the EPR sample at resonance.
- Q_μ is the effect that arises from surface currents in high-conductivity samples, and is proportional to B_1^2.

For a critically coupled resonator, Q_r equals the Q effects of all features internal to the resonator and $1/Q_r$ equals the sum of all other terms on the right hand side of this equation, so it can be rewritten as:

$$\frac{0.5}{Q_L} = \frac{1}{Q_U} + \frac{1}{Q_\varepsilon} + \frac{1}{Q_\chi} + \frac{1}{Q_\mu}. \tag{7.13}$$

For low conductivity samples Q_μ can be ignored, and for small samples Q_χ can be ignored, so the equation can be rewritten as:

$$\frac{0.5}{Q_L} = \frac{1}{Q_U} + \frac{1}{Q_\varepsilon}. \tag{7.14}$$

This shows that the two main effects on Q are the materials of construction of the resonator and dielectric loss in samples placed in the resonator. For the empty resonator, $Q_U = 2Q_L$.

Using the equations for inductance, L, and resistance, R, in (7.5), the equation for loaded Q becomes (Rinard et al. 1999a):

$$Q_L = \frac{d}{8}\sqrt{2\omega_0\mu_0\sigma} \tag{7.15}$$

from which it is seen that the frequency dependence of Q is proportional to the diameter of the resonator (d) and to the square root of its resonant frequency.

Other things being equal, a larger resonator has a higher Q. If the length of the resonator is not changed, the resistance of the walls of the resonator increases linearly with the increase in radius, but L increases as the square of the radius. Recall that this equation for Q is applicable to a series resonant circuit.

An approximate expression for Q, which has heuristic value, is in terms of resonator volume and surface area times the skin depth (δ) (Alger 1968, p. 109):

$$Q \approx \frac{\text{volume of the cavity}}{\delta(\text{surface area of the cavity})}. \tag{7.16}$$

This expression seems logical, since dissipation of energy occurs in the surface of the cavity, and more energy can be stored in a larger resonator. From this expression it is also easy to see that a resonator made of a poorer conductor (e.g., brass or aluminum instead of silver or copper) will have a lower Q.

7.5 Measurement of Resonator Q

The measurement of resonator Q is crucial to many EPR studies. This section provides guidance on measuring Q. Various laboratories have different instrumentation for electrical engineering test and measurement, so several options are

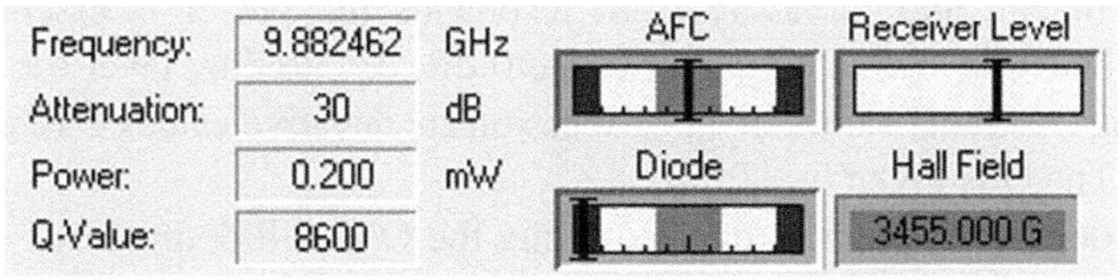

Fig. 7.3 The bridge indicators and Q-value reading from the Bruker Win Acquisition software

discussed. For readers who are not interested in the details of the measurements, the following section can be skipped.

7.5.1 Estimate Q Using the Bruker Software

The Bruker Xepr and Win Acquisition software provides an estimate of the resonator Q (Fig. 7.3), using the definition in terms of half-power bandwidth (7.3). If the Q is very low, or if the off-resonance baseline is not flat, the computer algorithm will not return an accurate estimate of Q. However, for the high-Q case the full reflection level is well-defined and the Q estimation is fairly accurate. In any case, the measurement is useful to reveal the similarity of samples for quantitative EPR and can be performed conveniently.

7.5.2 Q Measurement Using a Network Analyzer: By George A. Rinard

If a network analyzer, such as a HP8753D, is available the resonator Q can be measured accurately. Modern network analyzers have ways to calibrate the effect of transmission lines to the resonator. This may result in a measure of Q that is different from that measured for the resonator in a spectrometer. The reason for this is that when a resonator is "critically coupled" in a spectrometer, the operator really is coupling the resonator a bit away from critical coupling. The coupling and the frequency of the resonator are not those for minimum reflection from the resonator, but are offset slightly in order to provide a reflection from the resonator to null out leakage through the circulator. In essence, the circulator, transmission line, and resonator are being adjusted as a unit to minimize power at the detector at resonance.

When a resonator is critically coupled, negligible power is reflected at the resonant frequency, but when the resonator is overcoupled (or undercoupled) the power reflected at resonance is appreciable and another method is required to measure the loaded Q for these conditions. The following example is intended to clarify the method of measuring Q for the case of overcoupled resonators. The problem is that with a strongly overcoupled resonator the reflected power at resonance is not zero, so it must be determined in order to find the half power frequencies. The shape of the resonator dip is Lorentzian, and far away from resonance the power is essentially 100% reflected. This 100% reflected power is the 0 dB, or baseline, level for this method of measuring Q. All dB values in this discussion are negative values in reference to the 0 dB level. The Q is given by (7.17).

For any degree of under or over coupling the Q is still defined by

$$Q = \frac{\nu_0}{\nu_2 - \nu_1} \tag{7.17}$$

ν_0 is the resonant frequency and ν_1 and ν_2 are the lower and upper half-power frequencies. Measure the reflected power in dB at resonance (center of the dip), dB_0. The dB level for the two half-power frequencies is then given by (7.18).

$$dB_{hp} = 10\log\left(1 - \frac{1 - 10^{\frac{dB_0}{10}}}{2}\right) \tag{7.18}$$

It is convenient to calculate expected power levels with a programmable calculator, Mathcad or some other program. Example results are tabulated:

Reflected power at resonance, dB_0 (dB)	Reflected power at ν_1 and ν_2 (dB)
−2	−0.886
−20	−2.967
−40	−3.01

As expected, if the coupling results in −20 to −40 or more dB reflected at resonance, then the resonance curve is sharp enough that ν_1 and ν_2 are at 3 dB below the off-resonance reflected power.

7.5.3 Q by Ring Down Following a Pulse

Q can be measured on a pulsed EPR spectrometer. The power ring down following a pulse is a direct measure of resonator Q. Following a pulse, the power delivered by a cavity rings down exponentially:

$$U = U_0 e^{-\omega t/Q}. \tag{7.19}$$

The higher the Q, the longer the ring down time. This is why resonators are overcoupled to decrease the Q to minimize dead time in pulsed EPR experiments. Measuring Q of a very-low-Q resonator or of a strongly overcoupled resonator by ring-down is a bit more complicated, because miscellaneous reflections and switching transients in the system may distort the measured power from a pure exponential decay.

In this method, set up the spectrometer for a pulsed EPR measurement, which almost always involves overcoupling the resonator. Use a pulse that is long enough that the transient from the rising edge of the pulse has fully recovered before the end of the pulse. For example, at X-band a pulse length of 80–100 ns is adequate. Use a digital oscilloscope to record the shape of the reflected pulse, and fit the recovery following the end of the pulse to a single exponential. For example, on the Bruker E580, use a LeCroy oscilloscope as the transient recorder to record the shape using "random interleaved sampling" (RIS). Use Xepr to fit the exponential decay. The time constant for the decay is denoted here as τ. At X-band τ will be of the order of 2–4 ns for a strongly overcoupled resonator.

For a double-balanced mixer (DBM) biased on the LO side to, e.g., ca. 10 mW (+10 dBm), as in the Bruker pulse bridge, the output signals are linear in voltage, not power. Some mixers use different power levels, but the key point is that the LO power fully saturates the mixer diodes. For this case, the calculation of Q is:

$$Q_L = \pi \nu \tau, \tag{7.20}$$

where τ is the ring down time constant and ν is the resonator frequency. The ns and GHz cancel conveniently, so it is easy to keep track of orders of magnitude, and for the case of $\tau = 4$ ns at 9.5 GHz, Q = 120.

Home-built pulsed spectrometers often use a crystal to detect reflected pulse shape. For a crystal detector, with the power in the range where the output voltage is less than ca. 15–20 mV, the voltage output is linear in microwave power incident on the crystal. In this case Q can be calculated from the analogous ring down data using (7.21).

$$Q_L = 2\pi \nu \tau \tag{7.21}$$

Chapter 8
Filling Factor

8.1 General Definition

The filling factor (η) is another factor that is a linear contributor to the EPR signal voltage.

$$V_S = \chi'' \eta Q \sqrt{PZ_0} \tag{8.1}$$

As the name implies, the filling factor expresses the extent to which a sample fills the resonator. Since it is B_1 that creates the EPR signal, refinement of the concept of filling factor reflects the extent to which the sample occupies the region of the resonator in which B_1 is large. If B_1 were uniform in the resonator, the filling factor would be the ratio of the volume of the sample to the volume of the resonator. However, because B_1 is not uniform in a resonator, the calculation of filling factor is more complicated than a simple volume ratio. More precisely, it is the ratio of B_1 squared integrated over the sample volume to B_1 squared integrated over the resonator volume.

$$\eta = \frac{\int\limits_{sample} B_1^2 dV}{\int\limits_{cavity} B_1^2 dV} \tag{8.2}$$

For a small sample, the filling factor can be optimized by placing the sample at the maximum in B_1.

8.2 Calculation of Filling Factor

A good general rule is that a standard 4 mm o.d. quartz EPR tube, filled with a low-dielectric constant sample has roughly a 1% filling factor in an X-band TE_{102} cavity resonator. A loop gap resonator (LGR), a split ring, or a dielectric resonator sized for the same tube will have a filling factor that is tenfold higher than the TE_{102}

G.R. Eaton et al., *Quantitative EPR*,
DOI 10.1007/978-3-211-92948-3_8, © Springer-Verlag/Wien 2010

Table 8.1 Calculated filling factors

X-band resonator	Sample in 4 mm o.d. tube	Filling factor (%)
Bruker dielectric	2 mm o.d. by 10 mm long quartz rod	6.2
Bruker split ring	Same	9.5
Bruker rectangular cavity	Same	0.57

cavity resonator. However, less total sample resides in the LGR, split ring or dielectric resonator, because their active measuring region is much shorter than that of the TE_{102} cavity. The magnetic field modulation over the sample is usually more uniform in the smaller resonators because of their design. Detailed comparison of the net result on S/N requires finite element calculation of both the B_1 field and the modulation field.

Calculation of the filling factor requires accurate modeling of the variation of the microwave B_1 over the sample, which may be sample dependent. Modern finite element calculations, such as those performed with the Ansoft HFSS software, are required for accurate estimates of the filling factor for a sample and resonator combination. Table 8.1 presents some approximate calculations of filling factors, some of them performed with Ansoft HFSS. Since the user rarely has the opportunity to actually calculate the filling factor of a resonator, these values are intended to give the operator estimates of the magnitudes of filling factors for samples and resonators that approximate common experimental situations.

The filling factor for a sample with significant dielectric constant is increased by the effect of the dielectric in the microwave distribution (see Dalal et al. 1981a). Nagy (Nagy 1994) gave the following formula for the filling factor for a small spherical sample in a rectangular cavity.

$$\eta(\varepsilon', R) = \frac{16\pi R^3}{3V_{cavity}}\left(1 - \frac{k^2R^2}{5}\varepsilon'\right), \tag{8.3}$$

where R is the radius of the sample, and ε' is the dielectric constant of the sample.

Note that this formula predicts a decrease in η with an increase in either the radius or the dielectric constant, ε'. This is in contrast to the well-known increase in B_1 in a long cylindrical or flat sample in a TE_{102} cavity. Nagy (Nagy 1994) presented experimental data consistent with the prediction for a spherical sample. Subsequently, Yordanov and Slavov (1996) showed that the intensity of a point sample increases with an increase in the wall thickness of a long cylindrical quartz tube.

Attempting to achieve a high filling factor for a particular type of sample is a major incentive for creation of new types of resonators. Much of resonator design is concentrated on optimizing this ratio, the conversion efficiency, and minimizing the dielectric losses.

Chapter 9
Temperature

For many studies it is necessary to vary the temperature of the sample. Temperature control may be needed to hold a sample in a well defined state of equilibrium or to study at a series of temperatures. Cryogenic temperatures may be needed for transition metal ions with short relaxation times such that lines are too broad to detect at 298 K. To obtain quantitative results in variable temperature (VT) studies, sample preparation and careful temperature control are important.

9.1 Temperature Dependence of Signal Intensity

For substances that obey the Curie law, susceptibility is proportional to 1/T with T in K. This includes most S = 1/2 species except those with thermally accessible excited states. The relative population of two energy levels, such as the two energy levels of an electron spin in a magnetic field, is given by the Boltzmann distribution function. As is shown in standard texts, the ratio of spins in the high-energy state to those in the low-energy state is:

$$\frac{N_\alpha}{N_\beta} = e^{-\frac{g\beta B}{kT}}. \tag{9.1}$$

Since the population difference is very small, the difference in populations that is measured in EPR above 4 K, is accurately approximated as $\frac{g\beta B}{2kT}$. Since the energy level differences are very small, at X-band and above 4 K it is a reasonable approximation to say that the spin magnetic moment is

$$M = N\frac{g^2\beta^2 B}{4kT}. \tag{9.2}$$

Hence, relative intensities of signals that are not power saturated can be corrected for the Boltzmann population effect by multiplying by the ratio of the temperatures. The importance for quantitative EPR is that in variable temperature studies if all

DOI 10.1007/978-3-211-92948-3_9,

changes in Q, etc., have been accounted for, the EPR integrated intensity multiplied by the absolute temperature should be constant. This calculation is a good check to perform on VT EPR studies. Deviations could be due to violation of the assumption of independent spin behavior, or to partial saturation of some of the spectra.

Large changes in temperature also result in changes in the bulk magnetization for interacting spin systems. Crippa et al. (1971) reported the temperature dependence of the integrated EPR intensity for DPPH, Varian pitch, and lignite. The ratio of the areas depends on the temperature, because these samples exhibit different magnetic behaviors.

9.2 Sample Preparation for Cryogenic Temperatures

9.2.1 Selection of Solvent

Quantitative results at cryogen temperatures require that the sample does not precipitate or aggregate at low temperature. Formation of regions with locally high concentrations can be avoided by using solvents that form glasses rather than crystallizing. It may take some experimentation to find a solvent (usually a mixture) in which the species remains dissolved and which forms a glass upon cooling. Although there are many reports in the literature of spectra obtained in pure solvents, many such solvents including toluene and chloroform crystallize. Toluene may appear to form a good glass upon initial rapid freezing, but then crystallize during a series of variable temperature spectra. Many mixtures of common solvents reproducibly form glasses. Some convenient (and maybe non-obvious) solvent mixtures are listed in Drago (1977, p. 318) and Zecevic et al. (1998). Various mixtures of alcohols and branched-chain hydrocarbons tend to form glasses when cooled. However, ethanol undergoes a change at ca. −150°C that changes relaxation times. Various sugars and their derivatives form room-temperature glasses. Some solvents tend to break EPR tubes in certain temperature ranges. Water is extremely difficult to freeze and thaw in EPR tubes. 4:1 ethanol:glycerol often breaks tubes at about −120°C. Even 1:1 water:glycerol can be a problem because it undergoes a phase transition at about −100 to −90°C, and can break the sample tube if it is warmed slowly through this temperature. Triethanolamine forms good glasses when rapidly frozen, but if it is held at ca. 10°C the solution becomes cloudy. Medium-wall tubes are convenient for solvent mixtures that have a tendency to undergo phase transitions with volume changes (Table 9.1).

9.2.2 Sealed Samples

Samples for low-temperature spectroscopy often are flame sealed. However, a vacuum is an excellent insulator. Contraction of a sample when cooled can cause

Table 9.1 Solvents that form glasses[a]

1:1 anisole:3,4-di-methoxy-benzene
1:1:1 anisole: 3,4-di-methoxy-benzene: 2,5-di-methoxy-benzene
1:1 nBuOH:nPrOH
1:1 cyclotridecanone:cyclooctanone
decalin
decalin:3-Me-pentane (6:1 to 1:3 mixtures)
decalin:xylene (any ratio)
95:5 CH_2Cl_2:MeOH (Wada et al. 2006)
4:1 EtOH:MeOH
glycerol
6:1 to 1:3 glycerol:nPrOH
halocarbon oil $-(CF_2-CFCl)_n-$
1,3-bis-(1-naphthyl)-5-(2-napthyl)benzene (Swallen et al. 2006)
3-Me-pentane
2-Me-THF
2,5-Me_2THF
nujol
3:1 iso-octane:$tBuNH_2$
9:1 iPrOH:MeOH
9:1 nPrOH:MeOH
$(nPr)_2O$
sucrose octaacetate
o-terphenyl
o-terphenyl:decalin (mixtures from 19:1 to 1:1)
2:1 toluene:$CHCl_3$
9:1 toluene:THF
trehalose:sucrose (Dashnau et al. 2005)
9:1 trifluoroethanol:ethylene glycol
mixed tri-methylbenzenes
1:1 water:glycerol

[a]Based on experience in the Eaton laboratory, unless otherwise noted

a separation between the wall of the tube and the sample, which is then very well insulated from temperature changes. To facilitate temperature equilibration it is helpful to back-fill the tube with a low pressure of helium gas to provide heat transfer, prior to sealing the tube.

Caution. Sealed tubes stored in liquid nitrogen sometimes develop small cracks that permit the cryogen to get into the tube. When warmed to room temperature rapid vaporization of the liquid causes rapid expansion that can explode a tube. Therefore, it is very important to wear eye protection when handling sealed tubes.

9.3 Practical Aspects of Controlling and Measuring Sample Temperature

Cooling with liquid nitrogen or liquid helium is performed with two very different types of temperature control systems. They are available from various vendors, but for purposes of specificity the Oxford and Bruker products will be mentioned here.

9.3.1 Cavity Resonators

The operator needs to be aware that the temperature of the sample is not necessarily the temperature shown on the readout device. In gas-flow systems, there is a temperature gradient over the sample. Consider, for example, that just below the resonator the temperature is, say, 4.2 or 77 K, and a few cm above the resonator the top of the sample tube is at room temperature.

A measurement of the temperature gradient for two cases, −160 and 300°C, when using nitrogen gas flow over the sample was published in Varian 87-125-514 and is reproduced in Fig. 9.1. The gradients shown in the figure are exemplary only, and the actual gradient will depend on the gas-flow rate, the absolute temperature, the sample tube diameter and the Dewar inner diameter. For critical measurements, one should replace the sample tube with another tube of the same diameter that contains a thermocouple at the position of interest, in effect making the calibration of Fig. 9.1 for the resonator and gas flow system of interest.

The Oxford ESR900 style He gas continuous flow system (Fig. 9.2) cools the sample, analogous to the nitrogen flow systems used above 77 K. The temperature gradient across the sample can be rather large when using the ESR900. Our experience is that the actual temperature at the sample depends strongly on He gas flow (more sensitively than can be estimated from the flow meter) and on heater power. The temperature depends on exact dimensions of the sample tube, other things being constant. The temperature is also sensitive to the pressure in the liquid He Dewar, so it is a good idea to use the bladder provided by Oxford and a Bunsen valve to maintain constant Dewar pressure. For this style of cryostat, temperature may be the limiting uncertainty in quantitative EPR. Some of the newer transfer lines have larger diameter capillary tubes, which facilitate using N_2 as well as He, since N_2 is more viscous.

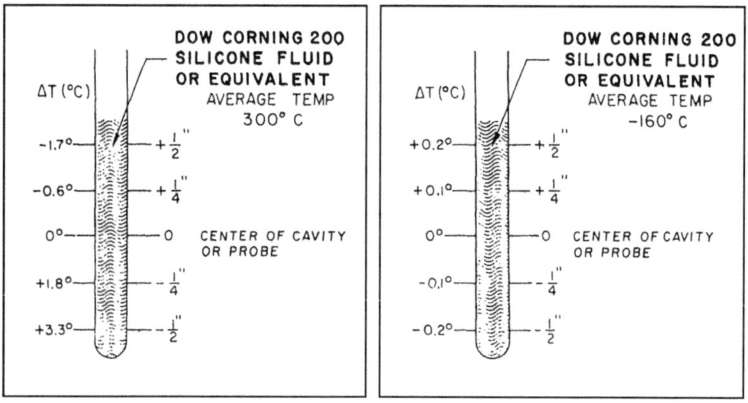

Fig. 9.1 Example of the temperature gradient for two cases, −160 and 300°C, when using a nitrogen gas flow over the sample in a TE_{102} cavity with a quartz Dewar insert. The figure is from the documentation for the E-257/WL-257 variable temperature accessory. Varian 87-125-514 B671

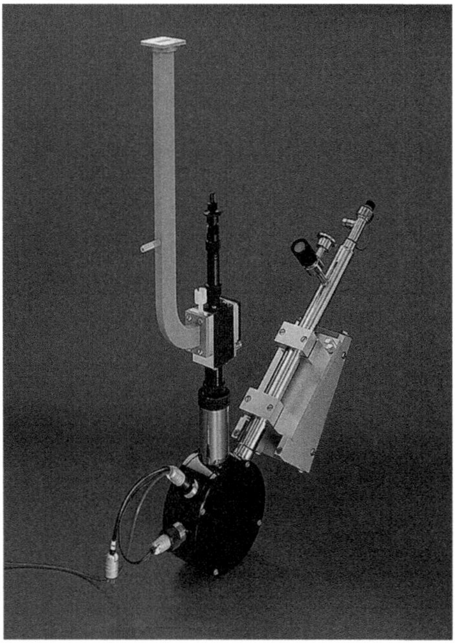

Fig. 9.2 Picture of ESR900 cryostat

Some EPR measurements near room temperature, such as of spin labels in membranes, require much more uniform temperature over the sample. For this, the temperature of the cavity, and not just the sample, has to be controlled. Alaouie and Smirnov (2006) published a method combining fluid-exchange temperature blocks mounted on the walls of a cavity and insulated from the magnet (see Fig. 9.3), with which they demonstrated ±0.033 K temperature variation over the sample length.

9.3.2 Flexline Resonators

The CF935 cryostat (Fig. 9.4) is designed for Bruker FlexLine resonators. In these cases, the cold gas cools both the resonator and the sample. There is an OFHP Cu region in the cryostat to create a uniform temperature region, and the temperature gradient over the sample can be negligibly small. However, it may take ca. 15 min for the sample temperature to become the same as the temperature of the sensor used in the feedback control circuit for the heater. Ideally, one would have a second temperature sensor, such as a Cernox, very close to or slightly above the sample, and wait until the two sensors agree before taking EPR spectra. With two calibrated sensors, one near and one below the sample, the temperature uncertainty is reduced to the calibration errors of the sensor. This is a large improvement over the ESR900. However, cooling a metal resonator to 77 K decreases the resistance of the metal, and increases the Q of the resonator a factor of 2–3 (Wright et al. 2000, and our

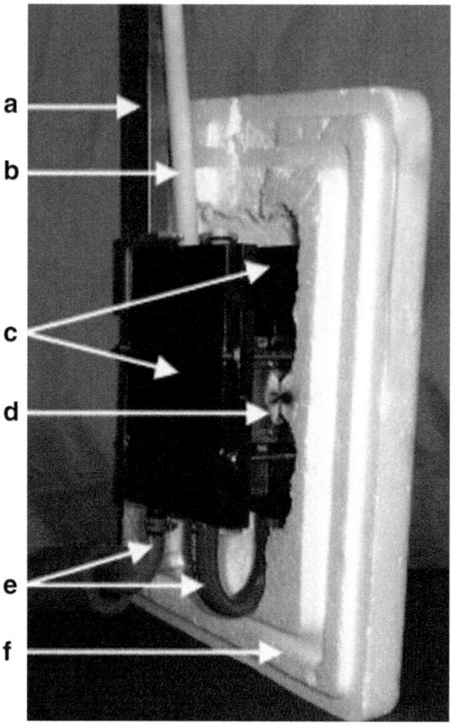

Fig. 9.3 Photograph of the NCSU VT probehead assembly for a Varian X-band TE102-type resonator with half of the Styrofoam insulation removed: (**a**) X-band waveguide, (**b**) iris tuning rod, (**c**) aluminum radiators, (**d**) front plate of the EPR resonator, (**e**) connecting tubing for heating/cooling fluid, (**f**) one of the two Styrofoam insulating enclosures. From Alaouie and Smirnov (2006)

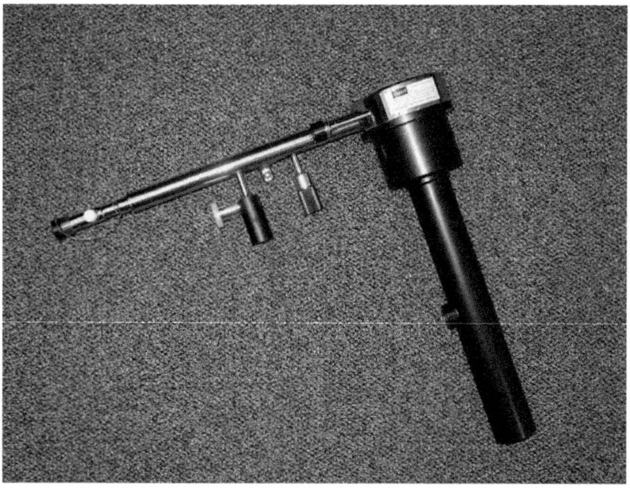

Fig. 9.4 Bruker/Oxford CF935 continuous flow cryostat for liquid He and liquid N_2 operation

observations). The temperature dependence of Q must be taken into account in quantitative measurements.

9.3.3 *Other Components of the Cooling Systems*

In addition to the two types of cryostats shown in Figs. 9.2 and 9.4, the full variable temperature system consists of the storage Dewar, a high-efficiency transfer line, a gas-flow monitor, and a diaphragm pump to pull the gas from the system and expel it either to a recovery system or to the atmosphere. The transfer line and ESR900 or CF935 Dewar should be evacuated before each use. In labs near sea level, it is usually adequate to pump with a diffusion pump or a turbomolecular pump and close the evacuation port valve prior to use. Because of the construction of the valves, it has been found necessary at high altitude (such as Denver and Colorado Springs) to continuously pump the cryostat during operation. Bruker has integrated a roughing pump and a turbomolecular pump into a convenient unit for evacuating transfer lines and cryostats (ER4112HV-PMT).

The cryogen flows through a very fine capillary in the transfer line. Humid air in the lab could condense in the capillary and block gas flow. Thus, it is good practice to purge the transfer line and the cryostat with dry gas (He gas is convenient) for a few minutes before inserting the transfer line into the cryogen storage Dewar or the EPR cryostat, and after use, while the transfer line warms to room temperature.

It is extremely easy to set up oscillations in He vapor, resulting in large temperature oscillations. Mechanical vibrations, such as someone bumping into the storage Dewar or the transfer line can start the oscillations, as can a too-rapid change in a flow-valve setting. When using He, it is best to throttle the liquid flow at the storage Dewar rather than gas flow at the input to the diaphragm pump. To stop the oscillations, temporarily cut back on the He flow and/or place the heater in manual mode rather than automatic control mode.

Cooling with liquid N_2 can be more difficult than cooling with liquid He. Liquid He is cold enough to cryopump imperfectly evacuated transfer lines, but with liquid N_2 residual gases in the vacuum shield can cause substantial heat transfer to the outer walls, resulting in very inefficient cryogen transfer. If the outside of the transfer line gets cold, reevacuate the line. Another problem is that N_2 freezes at slightly reduced pressure, so it is easy to block the capillary with frozen N_2 if the differential pressure is too large. Thus, for cooling with N_2, one throttles the flow before the diaphragm pump, rather than at the liquid storage tank, the opposite of the best practice for liquid He.

To avoid power saturation when performing measurements as a function of temperature, pay close attention to changes in T_1 and T_2 (Eaton and Eaton 2000b). As a guide, remember that T_1 of a nitroxide radical is ca. 200 μs at 100 K. T_1 varies approximately as T^2 in the temperature range commonly studied by chemists.

9.4 Operation Above Room Temperature

Operation above ca. 300 K requires special attention to the materials of construction of the resonator. The Bruker FlexLine resonator assemblies are designed for low-temperature operation, but not for operation above ca. 320 K. Operation with sample temperature controlled up to about 400 K can be performed with cavity resonators with internal Dewar inserts and heated N_2 gas. Higher temperatures require special resonators, such as the Bruker TE_{011} ER4114HT high temperature resonator for 400–1,200 K.

9.5 Example for S > 1/2

A bimodal resonator was used to record the EPR of a dimetal cluster in ribonucleotide reductase with B_1 parallel and perpendicular to the Zeeman field (Pierce et al. 2003). With CuEDTA as a standard to relate spin concentration to intensity of spectra, three species were followed quantitatively during titration with Mn(II). Simulations of X-band and Q-band spectra in perpendicular and parallel modes were used to quantitatively estimate concentrations of metal species. Since these samples involved spin-spin interaction, the signal intensities did not follow the Curie law. Figure 9.5 shows that the g = 11.7 and g = 4.6 signals in the $Mn_2^{II}R2$

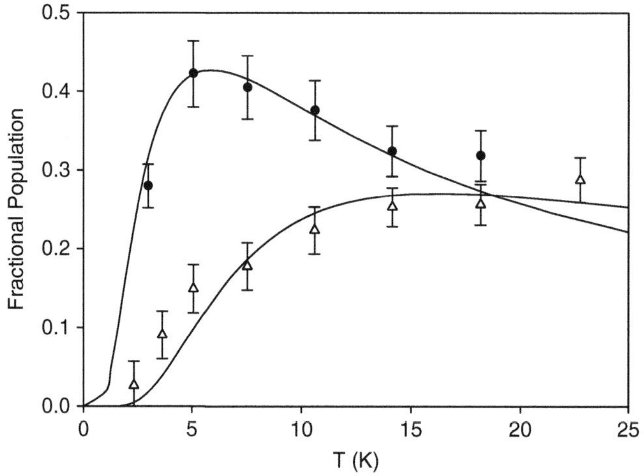

Fig. 9.5 Signal times temperature vs. temperature of the g = 11.7 (*black circle*) and g = 4.6 (*triangle*) signals in $Mn_2^{II}R2$. The solid lines are the theoretical populations of the S = 1 and S = 2 manifolds for $J = -1.8 \pm 0.3\ cm^{-1}$. The signal intensities times temperature (T) would be horizontal lines if the spin system followed the Curie law (intensity proportional to 1/T). Figure from Pierce et al. (2003)

samples have different temperature dependences, indicating that they come from different spin manifolds and that the two Mn(II) are antiferromagnetically coupled.

Another quantitative EPR study of a high-spin system provided the full energy matrix for $S = 7/2$ and showed excited state spectra that had maximum amplitude in the 20–50 K range (Hagen et al. 1987).

Chapter 10
Magnetic Field and Microwave Frequency

10.1 g-Factors

To measure g-factors the microwave frequency and the magnetic field must be determined precisely. Alternatively, a measurement can be made relative to a standard that is close enough in field that there is little error in the field offset.

Recent improvements in molecular orbital calculations of spin Hamiltonian parameters make it increasingly valuable to compare experimental and computer-derived values, even for transition metal ions. For example, Drew et al. (2007) used DFT to calculate g values of thiomolybdenyl complexes. The importance of accurate experimental values was highlighted in Drew et al. in the discussion of the results for one compound for which the experimental g values were higher than both the calculated values and the experimental values for similar compounds. In this example, the calculations may have been more accurate than the experiment.

10.2 Measurement of Microwave Frequency

A frequency counter is crucial for g factor measurements. With a well-calibrated counter, the largest source of error in g-factor measurements is the magnetic field measurement. Frequency counters change gradually with time. For example, a high-quality commercial frequency counter has a time base aging rate of about 1 ppm per year. For the most accurate measurements of g factors, the frequency counter should be recently calibrated. Some papers report g factors to precisions that approach the accuracy of the microwave counter.

The resonant microwave frequency usually is changed by inserting the sample in the resonator. This is caused by the dielectric constant (also called the relative permittivity) and conductivity of the sample. The dielectric constants of samples are usually higher than that of air, so the frequency decreases upon inserting the sample. One way to understand this effect is to consider the formula for capacitance:

G.R. Eaton et al., *Quantitative EPR*,
DOI 10.1007/978-3-211-92948-3_10, © Springer-Verlag/Wien 2010

$C = \varepsilon_r\varepsilon_0 A/d$ where A is the area of the capacitor and d is the distance between the plates of the capacitor. If the dimensions remain the same, the capacitance increases as the dielectric constant (ε_r) increases. Now, combine this with an expression for the frequency of a resonator, $\omega = 1/\sqrt{(LC)}$ where L is the inductance and C is the capacitance of the circuit (the cavity or LGR in this case). Inserting the sample increases C so it decreases the frequency.

10.3 Magnetic Field

The Bruker Hall probe is sufficiently accurate that it can be used as the standard for measurement of the magnetic field at X-band. However, the Hall probe and sample are not in the same place in the magnetic field. The operator needs to calibrate this offset. The cryostat affects the field offset between the sample position and the Hall probe position, so the correction should be determined with the cryostat in place. The offset can be several gauss. Bruker Xepr software includes a function for calibrating this offset so that accurate fields and g factors can be read from the spectra. For this purpose it is convenient to use solid DPPH, whose g value is generally quoted as 2.0036 or 2.0037 ± 0.002 (Wertz and Bolton 1972, p. 465; Weil et al. 1994b, p. 558). One can also use a sample with precisely known g-factor such as perylene radical cation. At other microwave frequencies, a Mn(II) sample is often used. There have been many attempts to reproducibly prepare a Mn(II) sample with well-defined hyperfine splitting to use as a magnetic field scan standard. However, the Mn(II) hyperfine is so exquisitely sensitive to details of the environment that this has been difficult. Yordanov (1994) tabulated some of the values for Mn in CaO, SrO, and MgO. The lore of the field is that Mn(II) samples usually turn out not to be sufficiently reproducible lab-to-lab to make them useful as a magnetic field scan standard. Bruker provides a calibrated Mn(II) sample with g = 2.0011 and hyperfine coupling constant A = 86.23 G.

To calculate the magnetic field differences for species with two different g values, g_1 and g_2, use the relation between field and frequency:

$$hv = g\beta B \tag{10.1}$$

$$\beta = \frac{e\hbar}{2m_e} = \frac{(1.60218 \times 10^{-19}C)(1.05457 \times 10^{-34}Js)}{2(9.10939 \times 10^{-33}kg)}$$
$$= 9.2740 \times 10^{-24}CJs/kg \tag{10.2}$$

These units are the same as JT^{-1}, since C = As, and $T = kg\ s^{-2}\ A^{-1}$ Rearrange (10.1) to the form:

$$B = \frac{hv}{g\beta} \tag{10.3}$$

For a particular case of X-band spectra (assume, e.g., 9.274 GHz) with $g_1 = 2.0023$ and $g_2 = 2.0123$ the resonant fields are:

$$B_1 = \frac{(6.62608 \times 10^{-34} Js)(9 \times 10^9 s^{-1})}{(2.0023)(9.2740 \times 10^{-24} JT^{-1})} = 0.32115T \quad (10.4)$$

$$B_2 = \frac{(6.62608 \times 10^{-34} Js)(9 \times 10^9 s^{-1})}{(2.0123)(9.2740 \times 10^{-24} JT^{-1})} = 0.31955T \quad (10.5)$$

In the more common units of G (1 G = 10^{-4} T) $B_1 - B_2 = 16$ G.

10.4 Magnetic Field Homogeneity

Magnetic field homogeneity depends on the magnet diameter and spacing between the poles, so the homogeneous volume is greater for a 10-in. (25 cm) magnet than for a 6-in. magnet. The specification for a 10-in. Bruker magnet is 13 mG in a cylindrical volume with 15 mm diameter and 15 mm long. This homogeneity is achieved and specified only at one magnetic field, because it depends on the degree of saturation of the iron in the magnet pole faces. Since narrow EPR lines and high-resolution spin-spin splitting is usually observed only very close to $g = 2$, the resolution of the magnetic field is specified at ca. 3,400 G for an X-band EPR spectrometer. Magnetic field homogeneity is relevant only over the size of the sample, and at X-band the 3-cm wavelength results in most signal intensity coming from about 1 cm length of sample.

Hysteresis and homogeneity. It is best to saturate the magnet before performing sweeps for very precise measurements. This means that the operator should increase the magnetic field to close to the maximum value achievable with the magnet and power supply combination. For example, depending on the power supply, a Bruker 10 in. magnet might be set to ca. 14 kG. The field is then set to the value about which one wants to make accurate field scans.

A good test of the magnetic field homogeneity is to record the CW spectrum of a narrow line signal. A sample that is commonly used is 50 μM perylene (Aldrich 394475) in sulfuric acid in a 0.6 mm i.d. capillary tube. Oxygen is removed by bubbling with N_2. In a sealed tube this sample lasts a few months at room temperature, so it should be kept cold when not in use. The line widths should be ca. 50 mG. Increases in line width imply magnetic field inhomogeneity over the length of the sample.

Sweep speed is limited by the physics of electromagnets, and especially by the limitations of iron-core electromagnets. The Xepr software display shows the field position with a red background when the scan speed is too fast for accurate field setting, and hence there is an effect on recorded line positions. However, the error in line positions occurs before the green background turns red. About 81 ms conversion time is OK, but faster may cause problems with some wide field sweeps.

When a large magnetic field scan is completed, and the field returns to low field, it could take several seconds to settle to the low-field starting point. If the next scan is started without waiting for the field to settle, false peaks may occur at low field because the field is still decreasing although the software is attempting to increase it. Consequently, for signal averaging of spectra with sweep of thousands of gauss, it is necessary to input a waiting time between scans. The delay that is needed depends on scan width, magnet, and power supply.

10.5 Coupling Constants Vs. Hyperfine Splittings

Electron-nuclear hyperfine coupling constants are characteristics of the paramagnetic species and are independent of the magnetic field or frequency at which spectra are recorded. If the hyperfine coupling constant is a small fraction of B_0 then the hyperfine lines are equally spaced and can be read quite accurately from the experimental spectra. However, as the ratio of the hyperfine coupling to B_0 increases, analysis of the energy levels for a field-swept spectrum becomes more complicated and hyperfine lines are not equally spaced. This situation frequently occurs for metal hyperfine splittings at X-band and for organic radicals at lower frequencies such as L-band (1–2 GHz) or 250 MHz. The contributions to the energy levels that give rise to unequal spacings of hyperfine lines are called Breit-Rabi effects. Computer-simulations that include Breit-Rabi effects give more accurate coupling constants than can be measured directly from the spectra, especially at frequencies below X-band or for very large hyperfine couplings. An example is shown in Fig. 10.1. The nitrogen hyperfine coupling constant for the nitroxide radical in water is $\sim$16.5 G, which results in three approximately equally spaced lines at X-band. However at 250 MHz the spacings of the three lines are not equal because the hyperfine coupling is a significant fraction of the approximately 90 G center field. In addition to changes in spacings between allowed transitions, some transitions that are "forbidden" at X-band become partially allowed at lower frequencies, which makes spectra more complicated as shown in Fig. 10.2.

10.6 Achievable Accuracy and Precision: g Value and Hyperfine Splitting

The accuracy of the field positions in a spectrum depends on both the digitizer that controls the field steps of a scan and on the field controller hardware. Older Bruker spectrometers such as the ER200 and ESP300 could do 4,096 steps in the field axis, using 12 bit resolution. The Elexsys and EMX spectrometers also have 12 bit magnetic field resolution (4,096 steps). The EMX-Plus and EMX Micro have 24 bit magnetic field axis digital resolution. The accuracy of the field setting is determined by the Hall probe. Bruker Hall probes are individually calibrated with

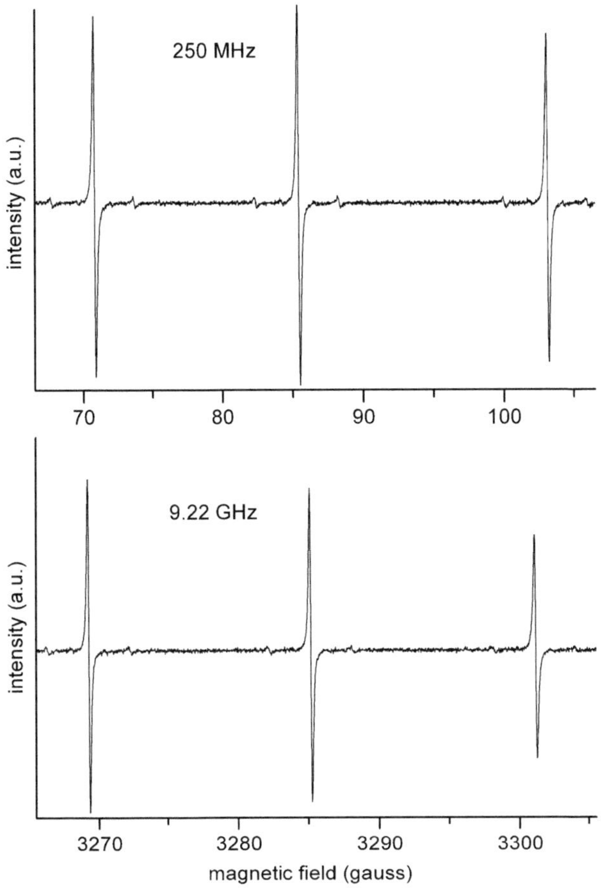

Fig. 10.1 Comparison of spectra of a nitroxide radical in water solution at 259 MHz (**a**) and X-band (**b**). Although the hyperfine coupling constant is independent of magnetic field, the spacings between the lines at low field are not equal due to the Breit-Rabi effect

a particular field controller. For older spectrometers the NMR Teslameter (also called a Gaussmeter) is somewhat more accurate than the field controller. When the spectrometer has been properly calibrated, either the field readout or the Teslameter provide equivalently accurate inputs for g-value and hyperfine measurements.

For the most accurate g value measurements, the operator needs to be aware of many subtle small effects, such as are well-described in Segal et al. (1965). For example, it was shown that the magnetic field inside a standard brass EPR cavity is not exactly reproducible when the field outside the cavity is reproduced exactly. A few g factors have been measured to high accuracy (see table in Weil et al. 1994b, p. 558; and Table 11.1). Awareness of the uncertainties of measuring both magnetic field and microwave frequency should caution against reporting a g factor to more than five significant figures, unless the spectrum of the species of interest was

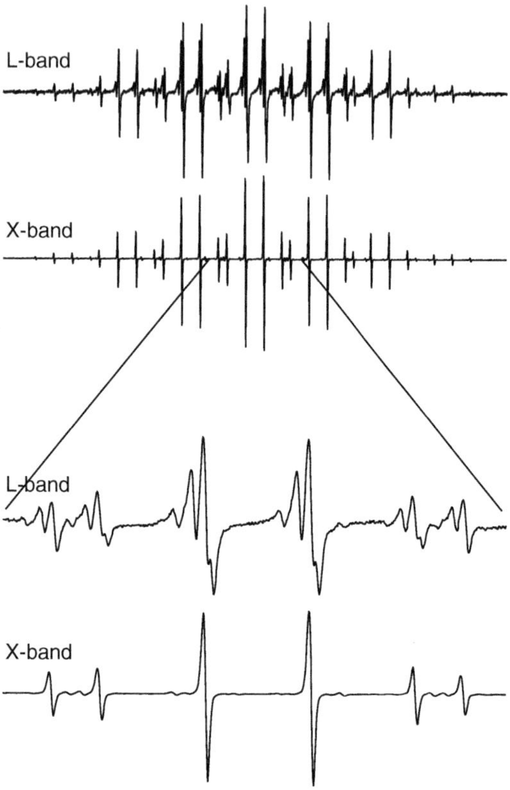

Fig. 10.2 Comparison of the spectra of PNT (perinaphthenyl) at L-band and X-band. Additional transitions that are forbidden at X-band become partially allowed at L-band

measured simultaneously with one of the radicals for which the g factor has been measured accurately. For routine g-factor measurements, it is handy to attach a speck of DPPH to the outside of a sample tube using tape such as "Scotch" brand transparent tape. Over a long period of time the DPPH partially dissolves in the adhesive and the signal broadens, but for the time of individual measurements this method works well enough to make g factor measurements to the accuracy of that of DPPH (± 0.0002).

Chapter 11
Standard Samples

In prior chapters many aspects of EPR spectroscopy that need to be considered to obtain quantitative results have been discussed. This chapter provides information about samples that have been proposed as standards. The standard and unknown should be as similar as possible. In all cases, it is important to pay attention to all of the parameters discussed in this book as being relevant to quantitative EPR. One anecdote may help one to remember this lesson. There is a paper in the literature, in which a small crystal of $CuSO_4{\cdot}5H_2O$, properly prepared, was used as an intensity standard to measure the number of spins in an aqueous protein sample at room temperature, without paying any attention to the differences in resonator Q or B_1 and modulation amplitude distributions for the two samples with very different geometries and dielectric loss properties.

11.1 Comparison with a Standard Sample

Absolute quantitation of EPR intensity is very difficult. In most cases, the spectrum of an unknown is compared with the spectrum of a known sample. Many samples proposed as standards for various purposes were discussed in Poole (1967, pp. 589–595), Wertz and Bolton (1972, pp. 462–466), Weil et al. (1994b, pp. 558–562), and Eaton and Eaton (1980). Note that weak pitch is a spectrometer performance standard, not an intensity standard. The derivative signal amplitude is calibrated under the stated spectrometer conditions for the "weak pitch S/N" test. The integrated area (total signal intensity) is not calibrated.

Extensive discussions of the challenge of performing quantitative EPR measurements included a focus on the issue of selecting an appropriate reference sample (Auteri et al. 1994; Czoch 1996; Dyrek et al. 1990, 1994, 1996; Nagy and Placek 1992; Nagy 1994; 1997; Siebert et al. 1994; Nagy et al. 1997a, b; Nagy and Sokolov 1997; Yordanov 1996; Yordanov and Christova 1997; Yordanov and Genova 1997; Yordanov et al. 1999).

G.R. Eaton et al., *Quantitative EPR*,
DOI 10.1007/978-3-211-92948-3_11, © Springer-Verlag/Wien 2010

Since DPPH has been used as a g-value marker, there has been much discussion of it as a standard also for signal intensity. However, the properties of DPPH depend strongly on the solvent from which it was crystallized (Yordanov 1996; Kolaczkowski et al. 1999), the purity of commercial samples varies, and the stability both as a solid and in solution is less than is needed for an intensity standard (Yordanov and Christova 1994, 1997; Yordanov and Genova 1997). Although DPPH is reasonably stable as the solid, in solution the stability is strongly dependent on the solvent, as reported by Slangen (1970), being much more stable in acetonitrile solution than in toluene solution. The solutions for these studies were not de-aerated and were stored in daylight, which impacts stability. Yordanov (1996) and Yordanov and Christova (1997) summarized prior literature on stability and molar absorptivity of DPPH and reported $\varepsilon = 12{,}350 \pm 650\ \mathrm{l\ M^{-1}\ cm^{-1}}$ at 520 nm in ethanol for use to determine the purity.

Nitroxide radicals can be obtained in sufficient purity that they can be used as standards for quantitating organic radicals in fluid and frozen solutions. Nitroxide radicals can be selected for solubility in the solvent of interest (ranging from toluene to water) so that the dielectric properties of the known and unknown samples can be matched. Commercial Tempol is purer and easier to handle than tempone, since it is a higher-melting crystalline solid, and is soluble in both water and alcohols.

K_3CrO_8 (Dalal et al. 1981b; Cage et al. 1998) has $g = 1.97$, which is far enough from many organic radical spectra that both can be measured at the same time, which is an aid to reducing uncertainties due to resonator Q, etc.

For transition metal samples, $S = 1/2$ Cu(II) and $S = 5/2$ Mn(II) are commonly used. $MnSO_4{\cdot}H_2O$ is available with 99% purity. However, it is somewhat efflorescent, so uncertainty in the degree of hydration could cause uncertainties in spin quantitation. Yordanov and Ivanova (1994b) documented some of the problems with Mn(II) standards. It might be more practical for EPR labs to purchase analytical standard solutions such as Alfa Specpure standards. Some labs try to use $CuSO_4{\cdot}5H_2O$ as a standard, but it is very difficult to prepare it and to store it in an atmosphere of the proper relative humidity to assure that it is really the pentahydrate. $CuCl_2{\cdot}2H_2O$ has also been cited as a primary standard, but its degree of hydration also depends strongly on the relative humidity of the air in which it is stored. For the most accurate work it is better to dissolve a weighed amount of Cu metal to use as a standard. For non-aqueous samples, compounds of the highest available purity should be used, and for the most accurate results they should be analyzed for metal content.

$VOSO_4{\cdot}nH_2O$ also has variable number of waters of hydration, so Dyrek et al. (1990) titrated with $KMnO_4$ in the usual volumetric analysis method to determine the V(IV) content. The compound was used as the solid, and in some cases ground with NaCl or KCl. Dyrek et al. (1994, 1996) found that the choice of diamagnetic diluent for a standard was not simple, and that the main problem is achieving a homogeneous distribution of paramagnetic component in the diamagnetic diluent. Grinding samples together to try to get a more homogeneous standard may result in chemical reactions that change the EPR spectra. The grinding process itself can produce radicals (Urbanski 1967; More et al. 1980; Adem et al. 1993; Tipikin et al. 1993, 1997). For example, Dyrek et al. (1996) observed that the $CuSO_4$ EPR spectrum changed when it was ground with NaCl, but not when ground with SiO_2 that had been pulverized before the two were ground together. Traces of water in the materials changed the spectrum of a $VOSO_4/K_2SO_4$ sample that was sealed in a quartz tube for a few weeks. The line shape changed, but the integrated intensity did not change. Drying over P_2O_5 resulted in samples that were stable for 5 years. In spite of sample preparation difficulties, Dyrek et al. (1994) concluded that preparation and measurement of the standard contributed only about 2% error to the overall quantitation of EPR spectra of metals.

For $S > 1/2$ it is important to take account of differences in transition probabilities as demonstrated by Siebert et al. (1994). Chromium in FeS_2 and in $AlCl_3{\cdot}6H_2O$ was measured by EPR and by ICPMS and AAS, with good agreement between methods. The National Bureau of Standards (now NIST) produced an EPR standard sample using Cr^{3+} in Al_2O_3 in 1978, which some labs still have (Standard Reference Material 2601 (Chang et al. 1978)). Nagy and coworkers discussed choosing reference samples for EPR concentration measurements (Nagy 1997; Nagy et al. 1997a, b; Nagy and Sokolov 1997). The focus was on transition intensity factors obtained by diagonalization of spin Hamiltonian matrices, in order to facilitate comparison of the intensity of e.g., one $S = 1$ species with the intensity of a different $S = 1$ center. Specific reference samples were not proposed in these papers.

Cordischi et al. (1999) compared a large series of compounds as possible primary standards for quantitative EPR of $S = 1/2$, 3/2, and 5/2 species. The $S = 1/2$ species (listed in Table 11.1) and $S = 5/2$ $MnSO_4{\cdot}H_2O$ were reliable standards, but "none of the pure Cr^{3+} compounds proved useful as primary standards because of their large fine structure terms or high Néel temperature that invalidated the simple Curie law." Among the Cr compounds tested were $KCr(SO_4)_2{\cdot}12H_2O$, $Cr(acac)_3$, and $(NH_4)_3[CrMo_6O_{24}H_6]{\cdot}7H_2O$.

As an internal standard for samples to be irradiated, Yordanov et al. (1999) suggest Mn(II) doped into MgO. They concluded that pyrolyzed sucrose and Mn(II) in CaO were less suitable. Irradiation of CaO induced EPR signals that could interfere with measurements of organic radicals, such as in alanine. There was no observable change in the intensity of the EPR signal in pyrolyzed sucrose upon irradiation, but the EPR signal of this material would overlap organic radical signals.

11.2 Spin Quantitation with a Calibrated Spectrometer

It is possible to determine the number of spins in a sample without the concurrent measurement of a reference sample. In this case, the double integrated intensity of the experimental spectrum and a software calculation are used to measure the number of spins. In order to perform the calculation, precise values for the resonator Q, microwave field distribution and modulation field distribution need to be known.

Table 11.1 EPR Intensity and g-Value Standards

Standard	Intensity	g-value	Hyperfine splitting	References
DPPH		2.0037 ± 0.0002		Weil and Anderson (1965)
DPPH	No	2.0036 ± 0.0001		Yordanov (1996)
DPPH	X			Cordischi et al. (1999)
Bruker or Varian strong pitch	±10%	2.0028		
Wurster's blue perchlorate		2.00305 ± 0.00002	Table of values	Randolph (1972a, b, pp. 98–101)
Perylene cation in 98% H_2SO_4		2.00258 ± 0.00002		Segal et al. (1965)
Tetracene cation in 98% H_2SO_4		2.00260 ± 0.00002		Segal et al. (1965)
p-Benzosemiquinone in butanol-KOH at 290K		2.004679 ± 0.0000129		Yordanov (1994)
Naphthalene anion radical		2.002757 ± 0.0000006		Yordanov (1994)
Fremy's salt $K_2NO(SO_3)_2$	X	2.0057	26.182 G between the outer lines	Randolph (1972a, b, p. 100) and Yordanov (1994)
Quinhydrone at pH 7.2	X			Narni et al. (1966)
K_3CrO_8	X	1.97		Dalal et al. (1981b), Cage et al. (1998)
$CuSO_4{\cdot}5H_2O$	X	$g_1 = 2.27$, $g_2 = 2.08$		Yordanov (1994)
$CuSO_4{\cdot}5H_2O$ solid, aqueous solution	X	$g_1 = 2.09$, $g_2 = 2.23$, $g_3 = 2.27$, $g = 2.18$		Cordischi et al. (1999)
$Cu(acac)_2$ solid or in toluene solution	X	$g = 2.13$, $g = 2.126$	A = 78 G	Cordischi et al. (1999)
Cu metal dissolved in acid	X			Eaton lab
Cr^{3+} in Al_2O_3	X			Chang et al. (1978)
Mn(II) doped into MgO	X			Yordanov et al. (1999)
$MnSO_4{\cdot}xH_2O$	X	2.0023		Yordanov (1994)
Tempo	X			Cordischi et al. (1999)
Tempone	X			
Tempol	X			
3-Trimethylamino-methyl-2,2,5,5-tetramethyl-1-pyrrolidinyloxyl iodide	X			Gerald Rosen
3-Carbamoyl-2,2,5,5-tetramethylpyrrolinyloxyl	X			
3-Carbamoyl-2,2,5,5-tetramethylpyrrolidinyloxyl	X			Towell and Kalyanaraman (1991)

$VOSO_4 \cdot nH_2O$	X			Dyrek et al. (1990)
$VOSO_4 \cdot 5H_2O$	X	g = 1.99		Cordischi et al. (1999)
$VO(acac)_2$	X	g = 2.00		Cordischi et al. (1999)
VOTPP (tetraphenyl porphyrin)	X	g = 1.958	A = 183 G	Cordischi et al. (1999)
Vanadyl and copper sulfate	X			Dyrek et al. (1994)
Fusinite	X			Auteri et al. (1994)
$MnSO_4$	X			Yordanov (1994)
$Cu(diethyldithiocarbamate)_2$	X			Yordanov (1994)
Pyrolyzed sucrose or dextrose	X	g = 2.0028		Yordanov (1994)
Ultramarine blue	X	g = 2.0294		Yordanov (1994)
F-centers in LiF	X			Yordanov (1994)
γ-Irradiated alanine	X			

Measurable parameters

A constant calculated from a sample with a known number of spins

$$DI = c \cdot [G_R \cdot C_t \cdot n] \cdot \left[\frac{\sqrt{P} \cdot B_m \cdot Q \cdot n_B \cdot S \cdot (S+1) \cdot n_S}{f(B_1, B_m)} \right]$$

Normalized spectrometer settings

Determined from EPR imaging

Fig. 11.1 An equation describing the double integral of an experimental EPR spectrum and how it relates to the number of spins in the sample

In addition, a constant (*c*) from a "one time" measurement of a reference spin standard is calculated. Once these values are input to the software's spin quantitation parameter file, the double integration of the experimental EPR spectrum is all that is needed to directly calculate the number of spins in a sample.

Figure 11.1 shows the various parameters that will affect the double integrated amplitude of an experimental EPR spectrum. These are input to the Quantitation parameter file for the software calculation. Where:

c = A constant input to the software from a sample with known number of spins
G_R = Receiver gain
C_t = Conversion time
n = Number of scans
P = Microwave power (W)
B_m = Modulation amplitude (Gauss)
Q = Quality factor of resonator
n_B = Boltzmann factor for temperature dependence
S = Total electron spin
n_s = Number of spins
$f(B_1, B_m)$ = Spatial distribution of the microwave field and the modulation field experienced by the sample.

The constant *c* is a value for the number of spins in a known reference standard. This value comes from a "one time measurement" of the reference sample and is then input to the software for the subsequent calculations where a reference standard does not need to be measured.

The parameters G_R, C_t, and *n* (receiver gain, conversion time and number of scans) are settings in the spectrometer's parameter file that can be normalized between different experimental spectra. The Q value is calculated by the software and also entered into the spin quantitation file.

Many of the values in the numerator of the second part of the equation (e.g., microwave power, and modulation amplitude) are available directly from the spectrometer's parameter file or can be input to the spin quantitation file (e.g., total spin).

The $f(B_1, B_m)$ term in the equation needs to be determined using a 3D EPR imaging experiment. This spatial distribution of the microwave and modulation fields

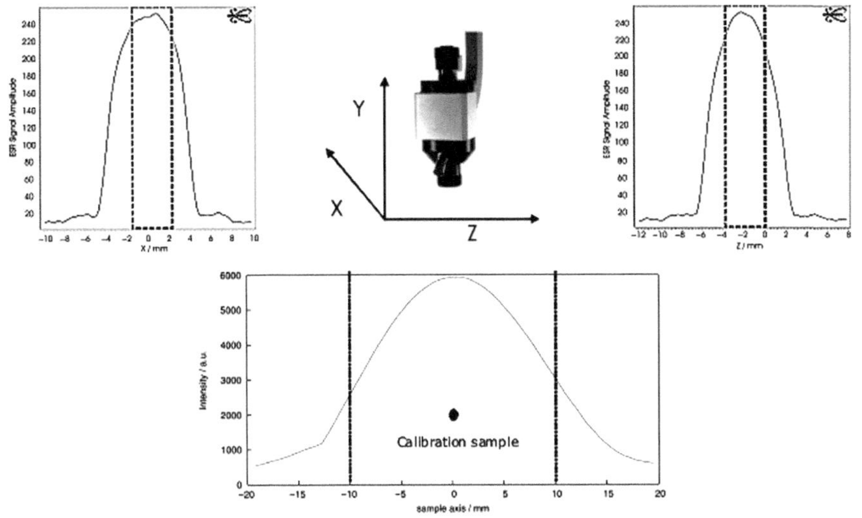

Fig. 11.2 Determination of the resonator profile (i.e., the $f(B_1, B_m)$ term) for a Bruker ER 4119HS resonator

is resonator specific and is determined by careful EPR imaging of a homogeneous sample that fills the entire volume of the resonator (see Fig. 11.2).

Obviously, the instrument and resonator specific parameters needed for the spin quantitation calculation are most easily determined by the manufacturer of the spectrometer and software. More detailed information on how this has been implemented in the Bruker Xenon software is available from Carl and Höfer (2008).

Appendix

Appendix A: Acquiring EPR Spectra and Optimizing Parameters

This appendix provides a step-by-step guide for obtaining an EPR spectrum and optimizing parameters, using Bruker software. The proper user-selected spectrometer settings (as discussed in detail in Chap. 4) are key to obtaining spectra that allow for accurate peak height measurements or double integrations. Selection of parameters for Tempol (4-hydroxy-2,2,6,6-tetramethylpiperidine-1-oxyl) is used as the example because it is a convenient reference standard for measurements involving various spin trap and spin label nitroxides. Remember that for quantitation the standard and the unknown should be dissolved in the same solvent, contained in the same size tube, and positioned similarly in the same resonator.

Measure the Spectrum with Nominal Settings

- Microwave power 5 mW
- Modulation amplitude 1 G
- Field sweep 200 G
- Scan time 5.1 s (if the sample is very weak, the scan time or gain may be increased) (Fig. A.1)

Optimize the Microwave Power

The microwave magnetic field (B_1) is what causes the EPR transition in the sample (see Chap. 6). B_1 is proportional to the square root of the microwave power output by the bridge. The "effective" intensity of B_1 over the sample depends on the

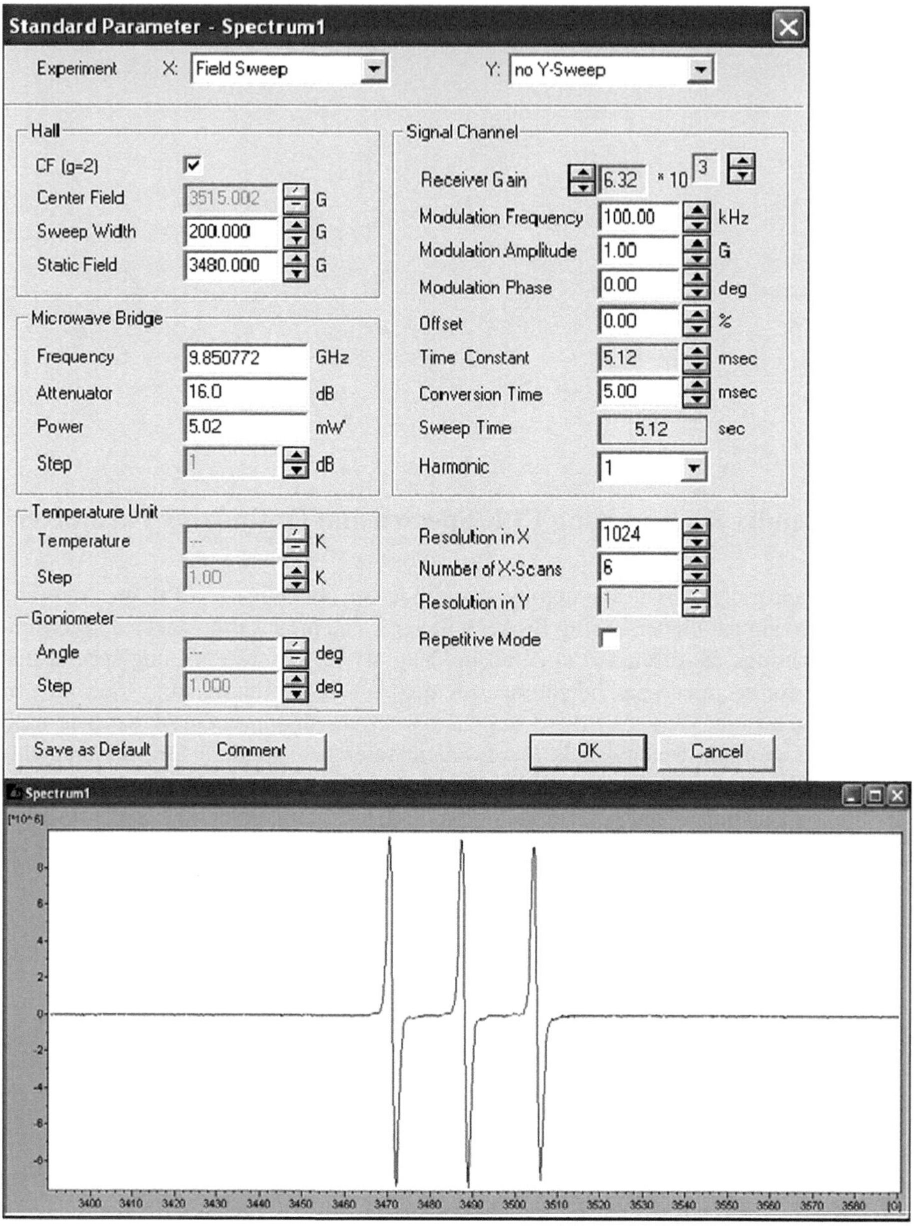

Fig. A.1 Measuring the initial EPR spectrum

incident microwave power, on the type of cavity or resonator that is used, and on the length of the sample. Therefore the power saturation behavior of a sample is determined experimentally. Figure A.2 shows the dependence on square root of the microwave power for the double integral (top curve), as well as the peak-

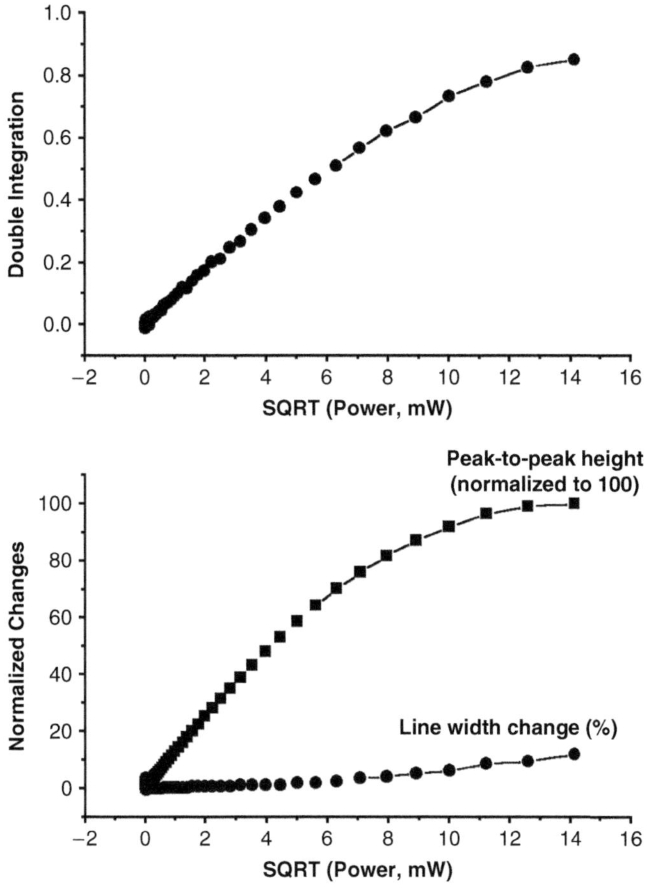

Fig. A.2 Effect of microwave power on the double intregral, peak-to-peak height, and linewidth of the nitroxide Tempol in water using an ER 4103TM cavity

to-peak height (a power saturation curve) and line width (bottom curves) for a rapidly tumbling Tempol sample. Most power saturation curves show a linear region (non-saturating) at low microwave power and a nonlinear region (saturating) at higher microwave power. In the region where the signal does not increase linearly with square root of power, the double integration increases nonlinearly and the linewidth increases. These data provide the basis for selecting instrument parameters for different experiments. For example, to simply maximize the signal intensity of a very weak sample for which power saturation is not of concern, a higher power level can be selected. However, if spectra will be used to characterize hyperfine splittings or line widths, a power in the non-saturating region should be selected. To quantitate the concentration of a sample using peak heights or double integrals, the spectra of both the sample and the spin standard should be measured under nonsaturating conditions. For Tempol in a TE_{102} cavity the maximum microwave power in the linear region is about four to five mW (Fig. A.2). Many organic radicals and spin adducts have power saturation curves in the same resonator that are similar to Tempol.

To select the appropriate power for a sample, first perform an automated 2D power sweep experiment, which records field sweeps at a series of microwave powers. The example in Fig. A.3 has a starting power of 200 mW (full leveled power) and successive scans at lower powers (i.e., 2 dB steps):

- Configure the power sweep experiment and collect the spectra
- Use the peak intensity (height) data to plot the power saturation curve
- Choose a microwave power level appropriate for the experiment
- To calculate spin concentration, choose a power level that is well within the linear region of the power saturation curve (e.g., $\sim$2 mW for the case shown in Fig. A.4).
- If the signals are weak and relative intensity measurements for samples with the same power saturation characteristics are adequate, a higher microwave power that provides improved signal-to-noise can be chosen.

Optimize the Modulation Amplitude

Field modulation amplitude also affects the line width and the peak-to-peak height. The data in Fig. A.5 show that as the modulation amplitude is increased, the line width remains constant to a point, after which it broadens.

The natural line width (Fig. A.6) of the signal can be estimated by measuring the spectrum with increasing modulation amplitude until the spectrum starts to broaden. Then make additional measurements at successively lower modulation amplitude values until the line width no longer gets smaller. The measured line width is then very close to the natural line width. If accurate lineshapes or resolution of small hyperfine splittings are important, the modulation amplitude should be less than 1/10 the natural linewidth. If the signal-to-noise is poor, it can be improved at the expense of linewidth accuracy (see Sect. 4.11). To maximize the signal-to-noise, the modulation amplitude can be set as high as about 1.5 times the natural line width. So, for example, if the natural line width is 1.7 G one might use 2.5 G modulation amplitude. The broadening of the signal by high modulation amplitude is called "over-modulation."

Figure A.7 shows the changes of peak-to-peak height of the Tempol signal with increasing field modulation amplitude. For Tempol the maximum peak-to-peak height is observed when the modulation amplitude is about 1.7 times the natural line width. Higher modulation amplitudes cause the peak height to decrease. Even for modulation amplitudes that are so large that the peak-to-peak height decreases, the double integration increases linearly with modulation amplitude (see Fig. A.8). This means that overmodulation of the signal can be used for quantitative comparisons of double integrals, provided that the field scan range is wide enough to accommodate the broadened spectrum.

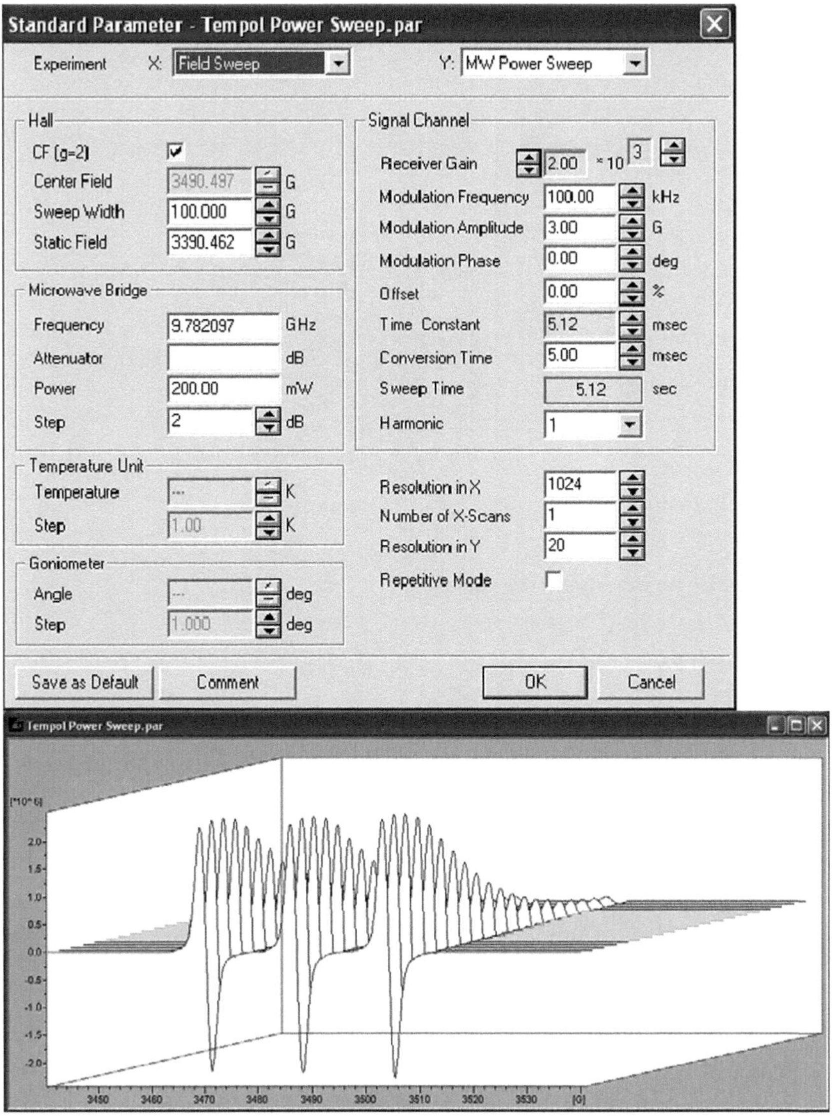

Fig. A.3 Parameters and spectra for the automated power sweep experiment

Optimize Magnetic Field Sweep Width and Number of Data Points

EPR lines often extend further into the outer regions of the spectrum than estimated at first glance. Therefore, it is important to use a magnetic field sweep width that is wide enough to encompass the entire spectrum. This effect is particularly important for the double integrations that are needed for signal quantitation. Figure A.9 shows

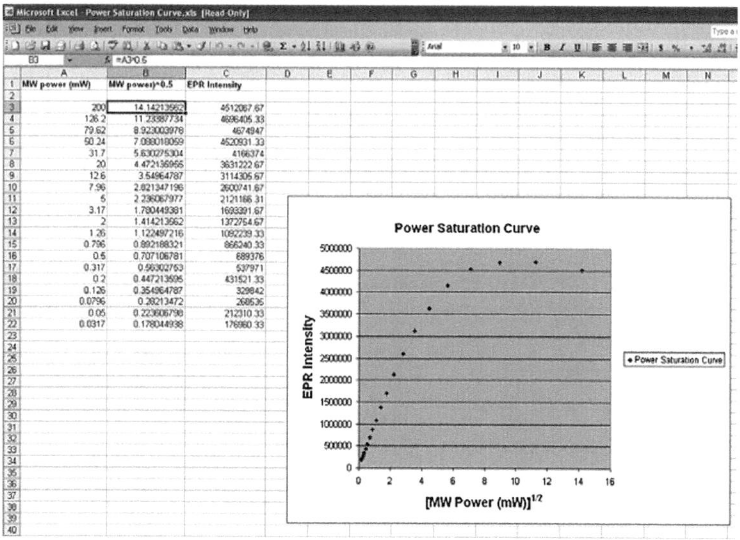

	MW power (mW)	MW power)^0.5	EPR Intensity
3	200	14.14213562	4512067.67
4	126.2	11.23387734	4696405.33
5	79.62	8.923003978	4674947
6	50.24	7.088018059	4520931.33
7	31.7	5.630275304	4166374
8	20	4.472135955	3631222.67
9	12.6	3.54964787	3114305.67
10	7.96	2.821347196	2600741.67
11	5	2.236067977	2121186.31
12	3.17	1.780449381	1693391.67
13	2	1.414213562	1372754.67
14	1.26	1.122497216	1092239.33
15	0.796	0.892188321	866240.33
16	0.5	0.707106781	689376
17	0.317	0.56302763	537971
18	0.2	0.447213595	431521.33
19	0.126	0.354964787	329842
20	0.0796	0.28213472	268536
21	0.05	0.223606798	212310.33
22	0.0317	0.178044938	176960.33

Fig. A.4 Plotting the power saturation curve

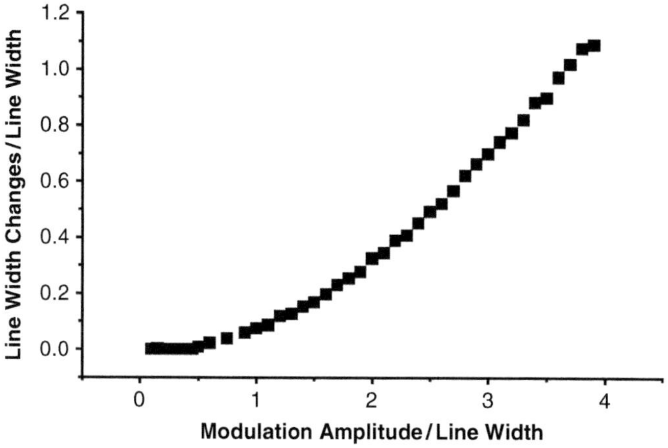

Fig. A.5 Line width changes with increasing field modulation amplitude for Tempol

the dependence of the double integral on sweep width. If the sweep width is not wide enough, a considerable portion of the spectrum is missing from the double integration. A good "rule of thumb" for Gaussian lines is to use a sweep width such that the distance between the first field value in the sweep and the start of the EPR signal is ten times the peak-to-peak linewidth of the narrowest signal (see "A value" in Fig. A.9).

So, for example, if the line width is 2.8 G, the field sweep should begin at least 28 G before the first line. If the line is Lorentzian, data extending for 30 linewidths

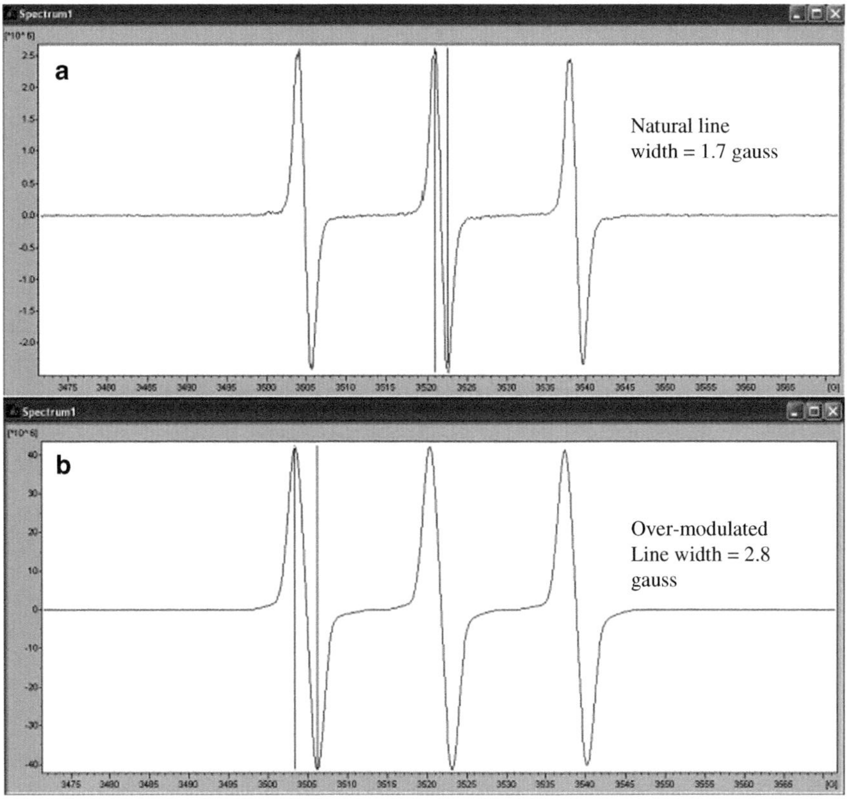

Fig. A.6 Measurement of the natural line width of the spectrum with a cursor tool in the acquisition software. Window A shows a modulation setting that provides a spectrum with the sample's natural line width. Window B shows a higher modulation amplitude setting and a slightly over-modulated spectrum

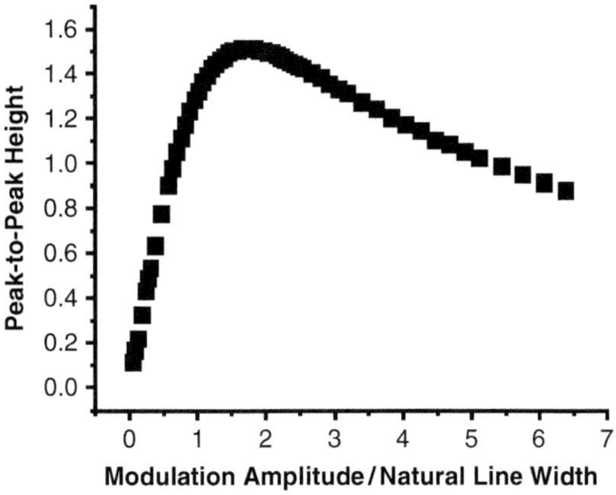

Fig. A.7 Changes in the peak-to-peak height of the Tempol signal with increasing modulation amplitude

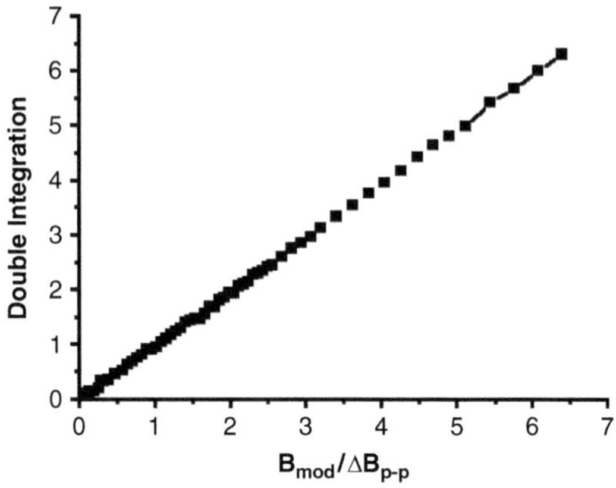

Fig. A.8 The double integral of the Tempol signal increases linearly with modulation amplitude

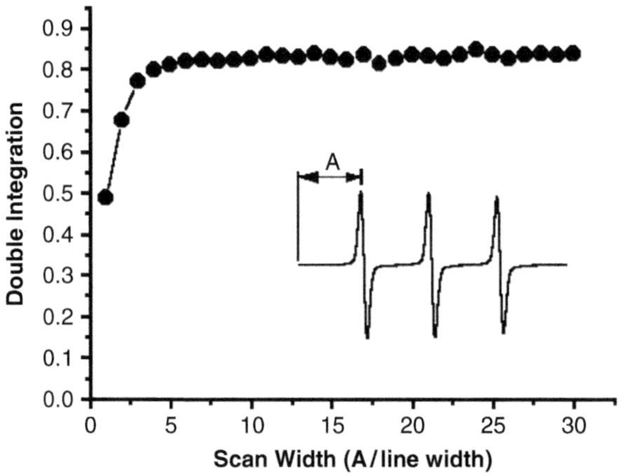

Fig. A.9 Double integral dependence on the field sweep width

is required for accurate integration (Fig. A.10). It is advisable to be conservative and perform a very wide sweep that assures all of the data from the EPR absorption is collected. Also, be sure that the number of data points is large enough that there are at least ten data points to define the peak-to-peak linewidth of the narrowest signal.

Summary

Configuring the spectrometer for quantitative experiments usually requires a compromise between settings that optimize the signal-to-noise, yet provide reproducible and fair comparisons between the reference standard and the unknown sample.

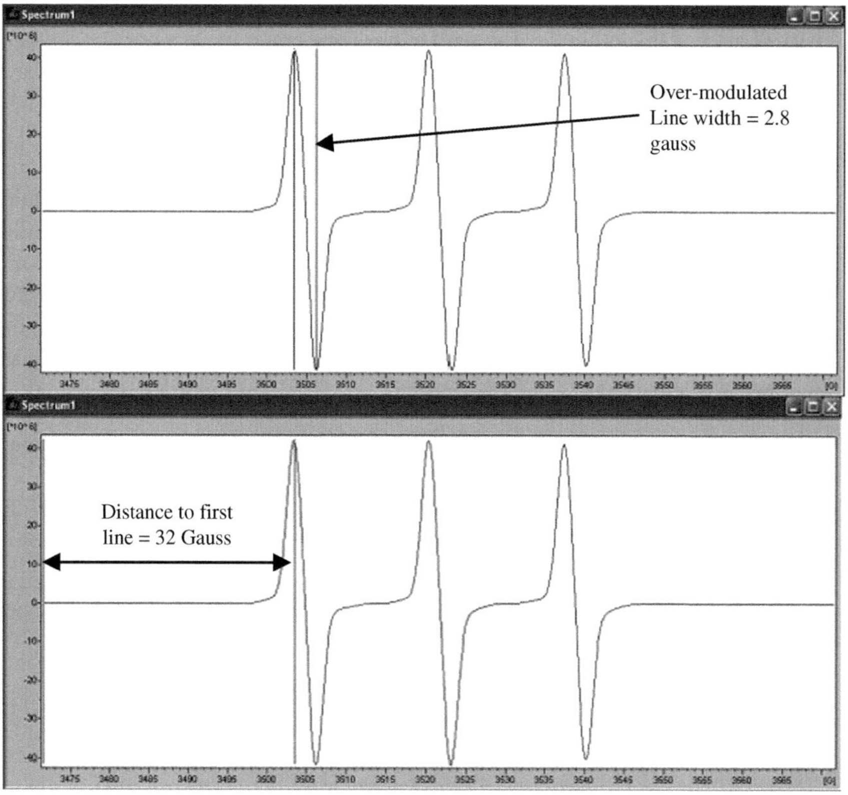

Fig. A.10 Measuring the line width and the distance from the first field point to the first EPR line

So, in summary, for a given amount of spectrum acquisition time, select values for the following parameters:

- *Microwave power* – to quantitate the spin concentration of a sample using the double integration method, the sample and the spin standard must be measured under nonsaturating conditions. Thus, a microwave power that is well within the linear region of the power saturation curve for the respective sample and standard must be used.
- *Modulation amplitude* – Since the double integration is proportional to the field modulation amplitude even if the signal is broadened, a relatively high field modulation can be used. In fact, aside from performing longer acquisitions. (e.g., performing long scans or signal averaging) increasing field modulation is the safest way to obtain better spectra of weak signal for double integration.
- *Sweep width* – Field sweep width is another factor that will adversely affect the double integration if set too low. The optimal sweep width depends on the line shape of the signal, but a safe rule of the thumb is to set the sweep width so that the distance from the field starting position to the first EPR line is ten times the

width of the narrowest line. Frequently the signal averaging method gives a better baseline than a single scan with a long time constant. In any case, it is best to keep the time constant and sweep time the same for the sample and standard. If a larger number of scans is needed for the unknown than for the standard, the double integrals can be divided by the number of scans averaged.

- *Maximizing signal intensity* – Remember, if the goal is only maximizing signal to noise (and not for quantitation), higher microwave power and field modulation amplitude can be used. However, if the microwave power or field modulation amplitude is too high the signal intensity may actually decrease. Set the field modulation amplitude at about 1.5 times the line width and the microwave power at the top of the power saturation curve to get the maximum signal intensity. This should give a good signal to noise enhancement although lines will be broadened.

Appendix B: Field Modulation and Phase Sensitive Detection

This appendix provides engineering level detail on how field modulation is used to detect the EPR signal. It can be omitted by readers who are not interested in this level of detail.

Details of Field Modulation and Phase Sensitive Detection

The EPR signal at microwave frequencies, such as 9.5 GHz, is "detected" by the crystal in the bridge. The output of the detector crystal is a DC (direct current) signal that is modulated as described in the following paragraph.

DC electronics are often noisy and the low-frequency components of the noise often appear as baseline instability. One common technique to remedy this problem is to modulate and then demodulate. By shifting a low frequency signal to a higher frequency (modulation), amplifying and processing it, and then converting it back into a low frequency signal (demodulation), superior baseline stability and noise reduction are obtained. An example of such a technique would be AM and FM radio. The audio signals are up converted to RF and then down converted to audio frequencies in the radio. An example in an EPR spectrometer is the use of field modulation. The microwave detector output from an EPR absorption is very noisy at low frequencies, but diminishes with 1/frequency as the frequency increases. The EPR signal is shifted to a higher frequency by modulating the magnetic field (often at 100 kHz). The EPR signal is then demodulated in the signal channel to obtain the desired noise-suppressed and stable low frequency EPR signal.

The heart of phase sensitive detection is a mixer. It has two inputs and one output. The output of the mixer is the product of the two input signals. The symbol for a mixer is shown in Fig. B.1. Notice the X which symbolizes multiplication.

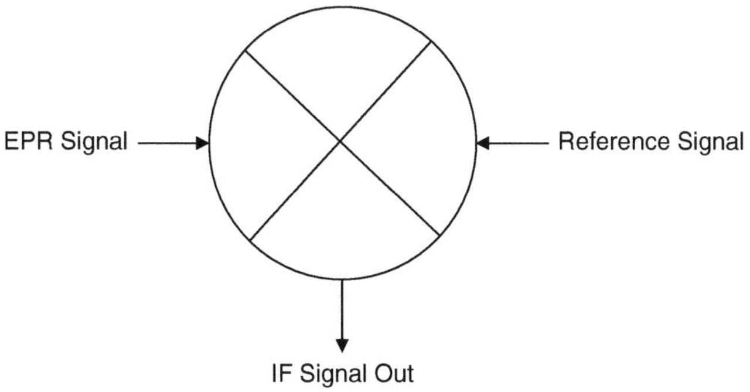

Fig. B.1 Schematic of a mixer

IF stands for Intermediate Frequency and is the resulting output signal that is subsequently filtered and processed to produce the EPR spectrum.

Modulation typically is sinusoidal. If a signal described by:

$$A_1 \cos(\omega_1 t + \varphi) \tag{B.1}$$

is applied to one input of the mixer and a signal described by:

$$A_2 \cos(\omega_2 t) \tag{B.2}$$

to the other input, the the product output signal is described by:

$$A_1 \cos(\omega_1 t + \varphi).A_2 \cos(\omega_2 t) \tag{B.3}$$

Using trigonometric relationships, this becomes:

$$\tfrac{1}{2} A_1 A_2 \{\cos((\omega_1 - \omega_2)t + \varphi) + \cos((\omega_1 + \omega_2)t + \varphi)\} \tag{B.4}$$

In this form three salient features of the IF output can be seen:

- The IF output can be decomposed into two frequency components: the sum and difference frequencies of the two input signals.
- The amplitude is equal to one half the product of the amplitudes of the two input signals.
- There is a phase factor, φ, equal to the phase difference between the two signals. This is the term that gives the name phase sensitive detection.

It is the sum and difference frequencies that make up and down conversion (or modulation and demodulation) possible. A high or low pass filter can be used to select either the sum or difference frequency component. The amplitude property facilitates the use of mixers for modulation and demodulation as well as feedback

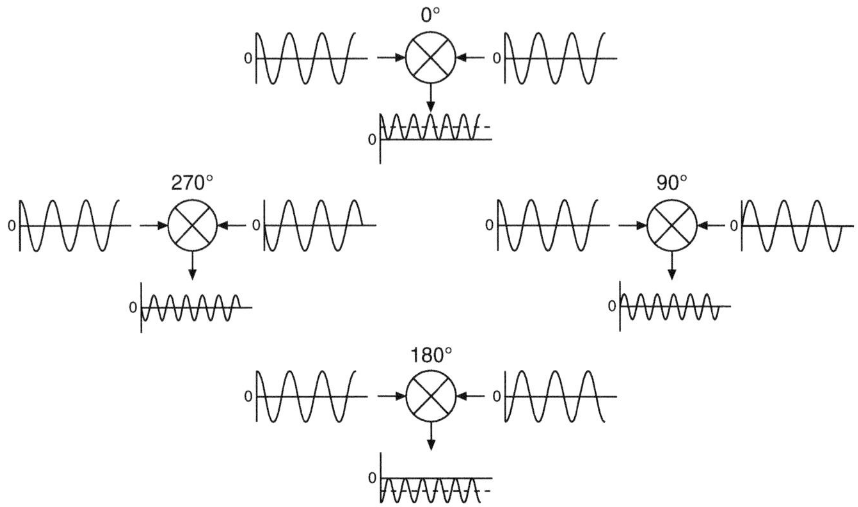

Fig. B.2 The IF output for different phase differences

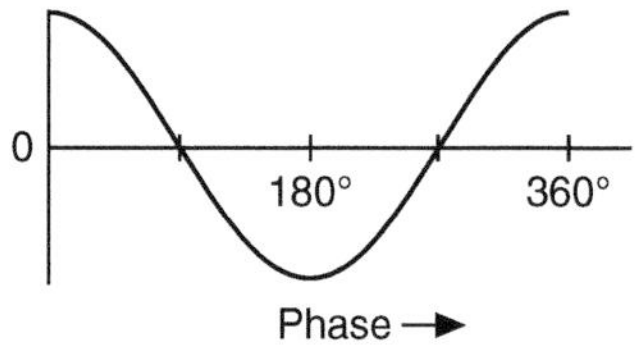

Fig. B.3 The output of a phase detector

loops. The phase factor can be used to discriminate between signals of different phase.

Quite often the LO (reference) input has the same frequency as the signal input and its amplitude is kept constant. The IF output is then zero frequency (DC) and double the frequency. Figure B.2 shows the IF output for various phase differences. The dashed lines indicate the DC components. Note the double frequency component as well.

If the low frequency component is selected by filtering out the high frequencies with a low pass filter, the IF output is proportional to:

$$A_2 \cos(\varphi) \qquad \text{(B.5)}$$

This scheme permits measurement of both the amplitude and phase of the input signal. If the input signal amplitudes remain constant, the IF output is proportional to cos(φ) and then the mixer functions as a phase detector (Fig. B.3). When the frequency difference and the phase difference between the two signals is zero, the DC output is at a maximum.

Field Modulation and Demodulation

The magnetic field at the sample can be modulated (often at 100 kHz) using the modulation coils (or posts) on the resonator. The resulting modulated signal is then demodulated in the signal channel. A block diagram of the circuitry in an EPR spectrometer for modulation and demodulation is shown in Fig. B.4. The relationship between the field modulated signal and the first derivative of the EPR absorption lineshape is depicted in Fig. B.5.

A Visual Description of Why the EPR Signal Appears in the First Derivative Form

The following figures are used to illustrate how the first derivative spectrum is the resulting output from the signal that is demodulated in the signal channel. The Modulated Signal (Fig. B.6) originates from the reflected microwaves coming from the cavity and is the input to the signal channel. The Unfiltered Demodulated Signal is the phase detected signal before it is filtered by the signal channel. The dashed horizontal line in that display is the average value of the unfiltered signal and is equivalent to the signal after it is filtered by the time constant. The two displays on the left side are the EPR absorption signal and the corresponding phase detected and filtered EPR signal. The vertical dashed line indicates the magnetic field value at a given time in the EPR experiment. The Demodulated Lorentzian display is the phase detected and filtered EPR signal plotted as a function of the magnetic field. In each successive figure, a black dot data point is added to the recorded output spectrum. These data points correspond to the position (or amplitude) of the

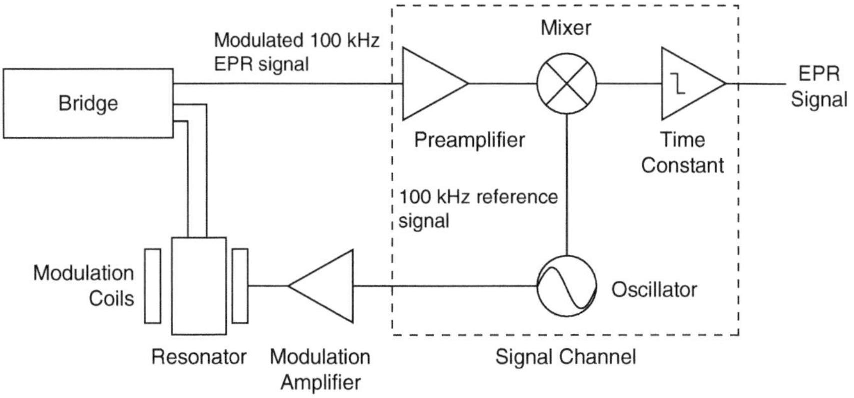

Fig. B.4 A simplified diagram showing parts of the EPR spectrometer that are used for field modulation and phase sensitive detection. The microwave EPR signal is converted to a modulated DC signal by the crystal detector in the bridge

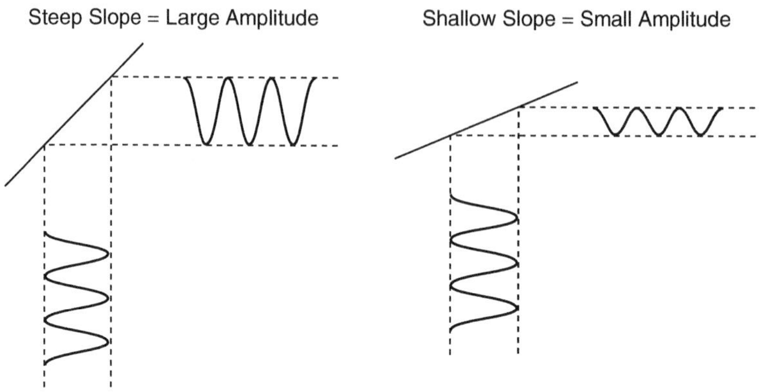

Fig. B.5 Conversion of modulation into a modulated signal

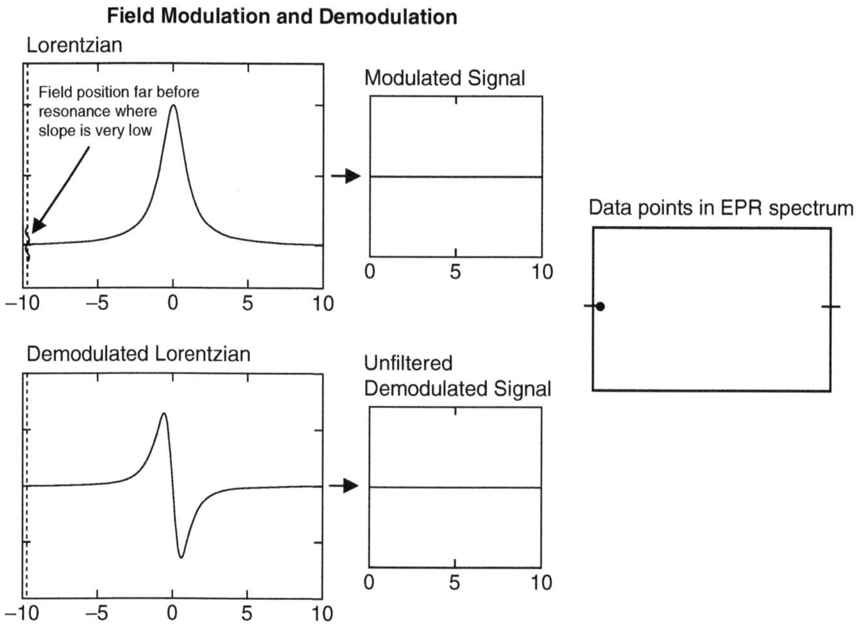

Fig. B.6 When the field is at a position far before resonance the slope is near zero and so a baseline data point is recorded

horizontal dashed line that is shown in the plot of the "Unfiltered Demodulated Signal."

If the magnetic field is moved to a position in the EPR signal, an oscillation is observed with amplitude that is approximately proportional to the slope of the signal (Fig. B.7).

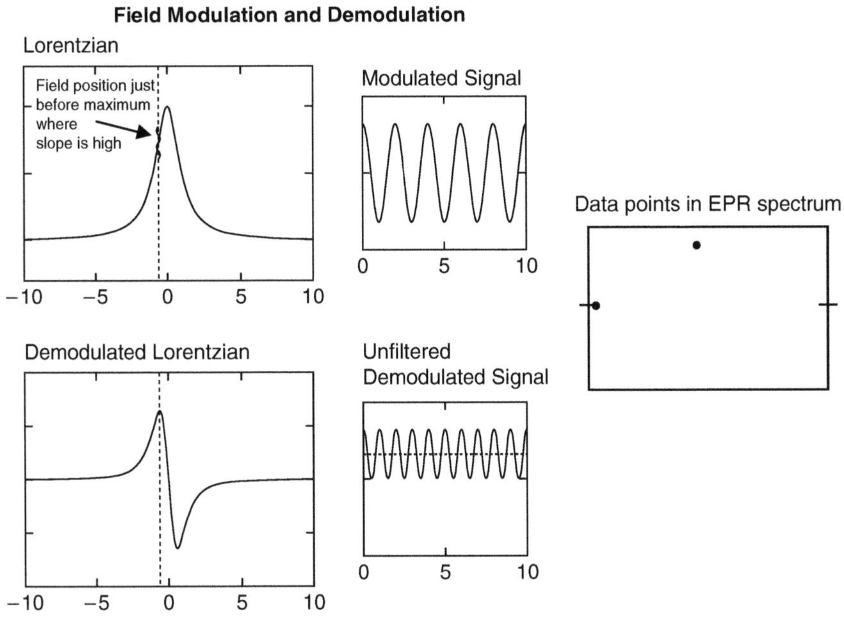

Fig. B.7 Approaching the center of the EPR absorption signal in a region with a high (positive) slope

At the top of the EPR absorption, the output of the mixer has twice the modulation frequency. This results from the fact that the EPR absorption signal is symmetric at that point. The filtered demodulated signal drops to zero here. This data point provides the "zero cross" or baseline value within the EPR absorption envelope (see Fig. B.8).

If the field is shifted further, a 180° phase shift in the modulated signal can be seen by comparing the modulated signals in Figs. B.7 and B.9. Also notice that the filtered demodulated signal (the horizontal dashed line) becomes negative after the midpoint of the EPR absorption.

And once again, for a magnetic field that is well past the field for resonance, the slope is very near zero corresponding to a baseline point (see Fig. B.10). For illustrative purposes, only five data points were shown while stepping through this example. Figure B.10 shows the actual spectrum that would be obtained with multiple points across the field scan range.

Suppression of 1/f Noise

Another reason for using modulation and phase sensitive detection is to suppress 1/f noise. Noise drops off at higher frequencies. If no modulation is used, the EPR signal has predominately low frequency components. The output of the diode detector is dominated by the 1/f noise and not by the EPR signal (see Fig. B.11a).

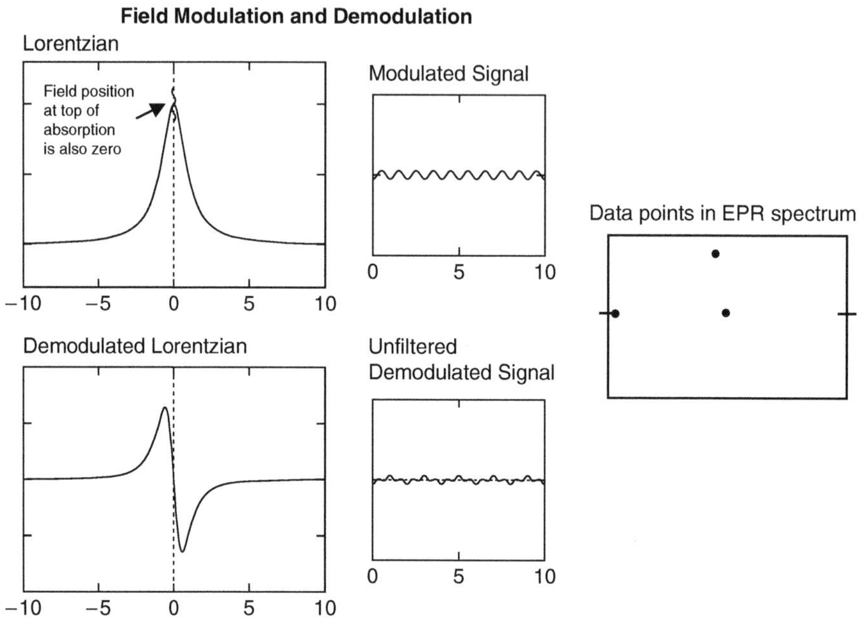

Fig. B.8 Signals at the center of the EPR absorption signal

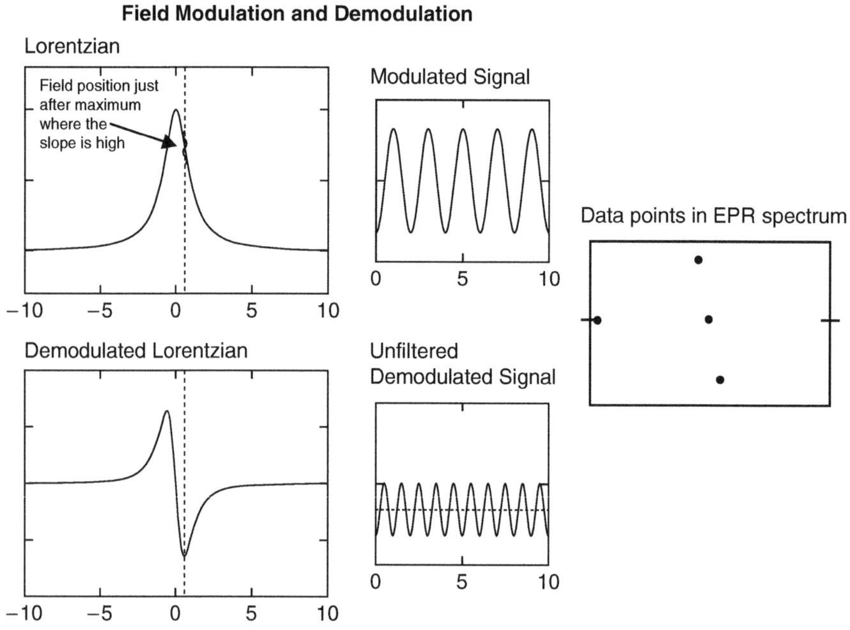

Fig. B.9 Signals on the right side of the EPR absorption, but still in a region with a high (negative) slope

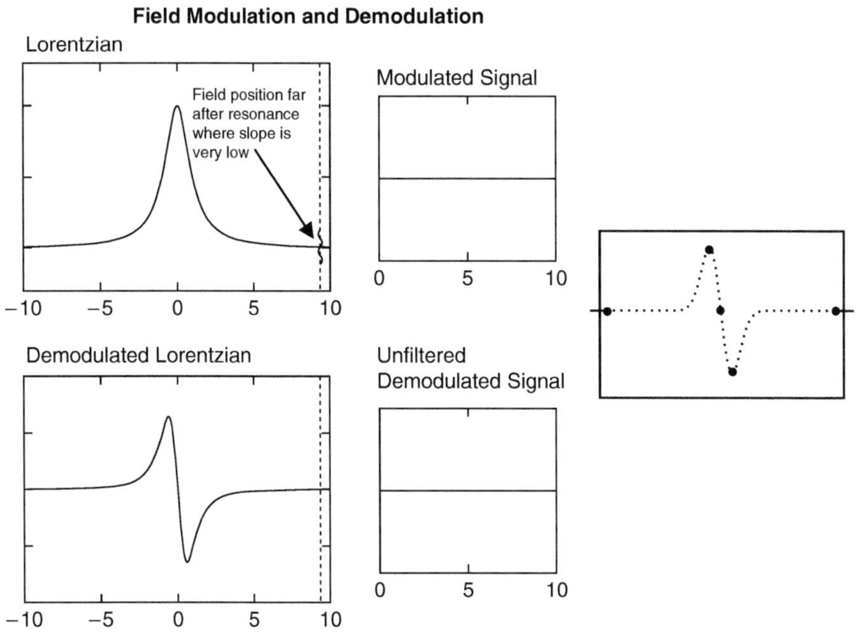

Fig. B.10 Signals for a magnetic field position that is well past resonance, and the lineshape that would be obtained if multiple additional points had been examined

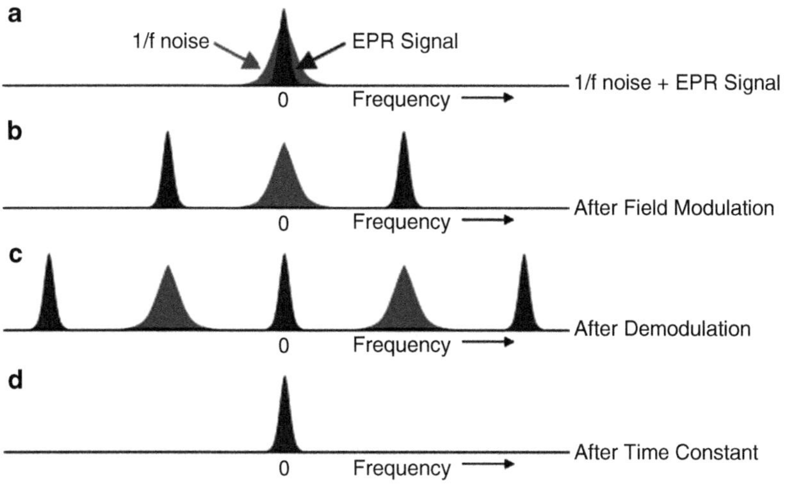

Fig. B.11 A cartoon of how field modulation and phase sensitive detection is realized in an EPR spectrometer

If 100 kHz field modulation is applied, the EPR signal is shifted to 100 kHz (see Fig. B.11b). The noise is not frequency shifted because the field modulation only has an effect on anything that is magnetic field dependent such as the EPR signal and not on anything field independent such as the 1/f noise. There is no mechanism for the field modulation to affect the 1/f noise.

The EPR signal is now nicely separated from the 1/f noise, but alas it is at 100 kHz. After phase sensitive detection, our EPR signal is shifted to 0 Hz and 200 kHz (see Fig. B.11c.). The 1/f noise is also shifted to 100 kHz. Unlike in the case of modulation having no effect on the noise, the 1/f noise, like any signal, will exhibit both sum and difference frequencies after demodulation. Finally, the unwanted 200 kHz EPR signal and 1/f noise are filtered out by the RC filter with a selectable time constant (see Fig. B.11d.), and the unwanted 1/f noise has been successfully suppressed.

Appendix C: Post Processing for Optimal Quantitative Results

A user may want to make relative intensity measurements with samples that only provide very noisy spectra or may want to perform double integrations on spectra with background or baseline problems. In such cases, there are a number of post processing procedures that can greatly improve the accuracy and precision of the quantitative EPR measurements. In this appendix two examples are outlined for improving experimental spectra through post acquisition processing.

Example of Baseline Subtraction to Improve Spectrum for Double Integration

Double integration of the first derivative EPR spectrum is used to quantitate EPR samples. Because most spectrometers record the EPR signal as a first derivative of the absorption signal, the spectrum must be integrated once to recover the absorption spectrum and then integrated a second time to obtain the area under the absorption curve. Spectral processing software performs the double integration. The following section demonstrates how to perform double integrations using Bruker's WIN EPR spectral processing software and also provides suggestions for improving the accuracy of double integrations. WIN EPR is used only as one example; there are many other software programs available that can perform the same tasks.

It is important to realize that even slight baseline drifts, background signals, or a very low signal to noise ratio can be detrimental to the accuracy of double integrations. Fortunately, there are many tell-tale signs that warn a user of a poor integration. Figure C.1 shows how a double integration was improved by signal averaging

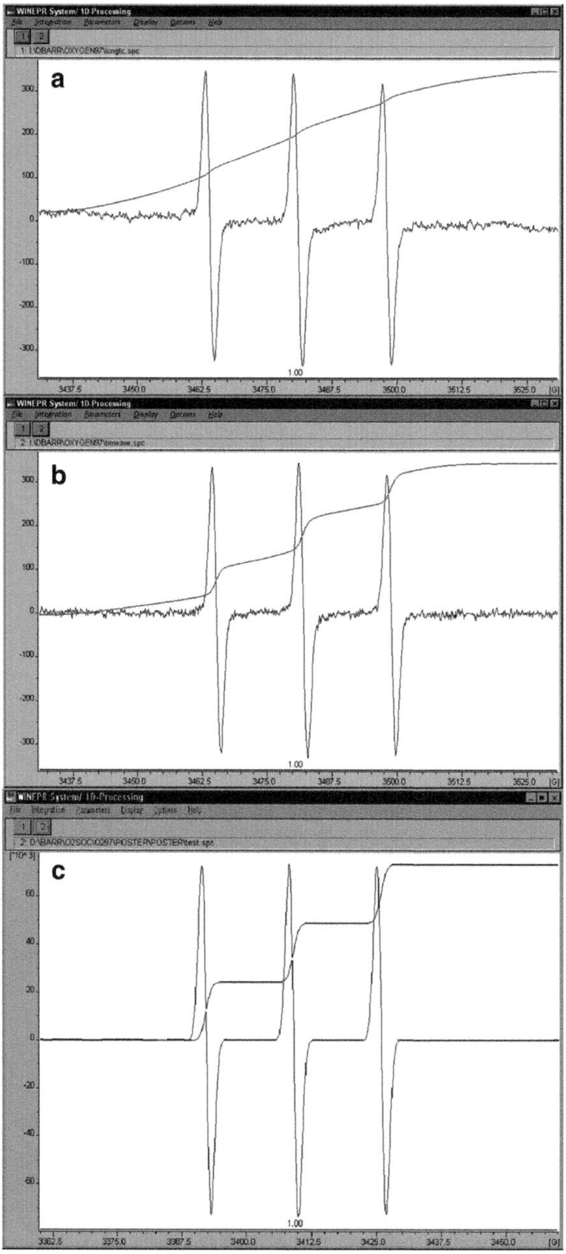

Fig. C.1 Spectrum (**a**) was collected using a 326 ms time constant and a 335 s scan. Spectrum (**b**) was recorded using a series of 32 scans with a 10.2 ms time constant

Note: The total acquisition times for (**a**) and (**b**) were the same. Spectrum (**c**) is a simulation of the experimental spectrum. Double integrations are overlaid on each spectrum

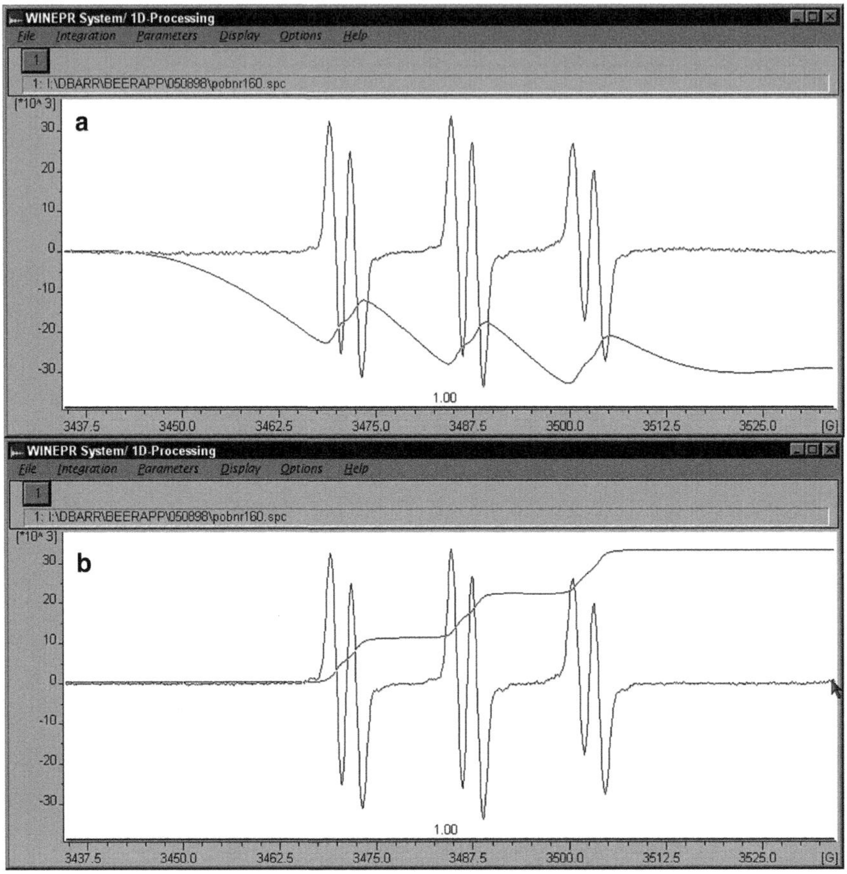

Fig. C.2 Improvement of double integration using the baseline fit and subtract feature of WIN EPR. Spectrum (**a**), is the first derivative spectrum before baseline correction. Spectrum (**b**), is the first derivative spectrum after subtracting the baseline. The double integrations appear overlaying each spectrum

a series of short scans with a short time constant, or by simulating the experimental spectrum. The double integrations are overlaid on each spectrum. In (**a**), the double integral curve starts rising long before the EPR signal begins. Although (**b**) is much improved, the curve still starts to rise before the EPR signal begins. Meanwhile, the double integration of the simulated spectrum (**c**) has no confounding baseline effects (i.e., only the EPR signal of interest has been calculated).

Seemingly insignificant baseline drifts and/or background signals often have a confounding effect on the double integration of the EPR spectrum. Consider a spectrum that also has an extremely broad background signal. Even though the peak-to-peak height of the background may be so small that it is unrecognizable, its double integrated intensity may be larger than a very narrow signal that dominates the first derivative representation of the spectrum. In Fig. C.2 spectra (**a**) and (**b**) are actually the same EPR spectrum, except in (**b**), the baseline has been subtracted. The double integrations are displayed overlaying both spectra. Although the first derivative representations seem almost identical, the double integration of the uncorrected spectrum in (**a**) is distorted. This would, undoubtedly, give a misleading result if it were used in a quantitative study. The following paragraphs demonstrate how post processing software can be used not only to obtain numerical double integration values, but also how to improve the accuracy of these values when the EPR spectrum has an unavoidable background signal or baseline drift.

Convert the First Derivative EPR Spectrum into an Absorption Spectrum

It is often easier to visualize and to correct a distorted baseline when working from the absorption spectrum. To convert the first derivative to an absorption curve, select "Integrate" under the "1D-Processing" menu (Fig. C.3).

Correct the Baseline

It becomes apparent from the absorption spectrum that the baseline is not flat (Fig. C.4). To calculate the baseline, first select "Baseline Correction..." under the "1D-Processing" drop-down menu. From the "Baseline Correction..." menu, select "Define Region" and define the region of the spectrum. Do this by clicking the left mouse button to designate the starting point of a region. Then drag the pointer to the desired end of a region and click the right mouse button. The regions that were

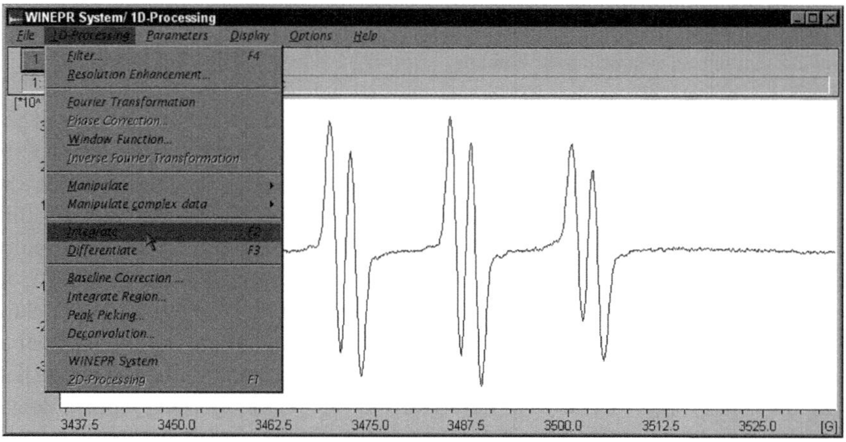

Fig. C.3 Converting the spectrum to an absorption representation

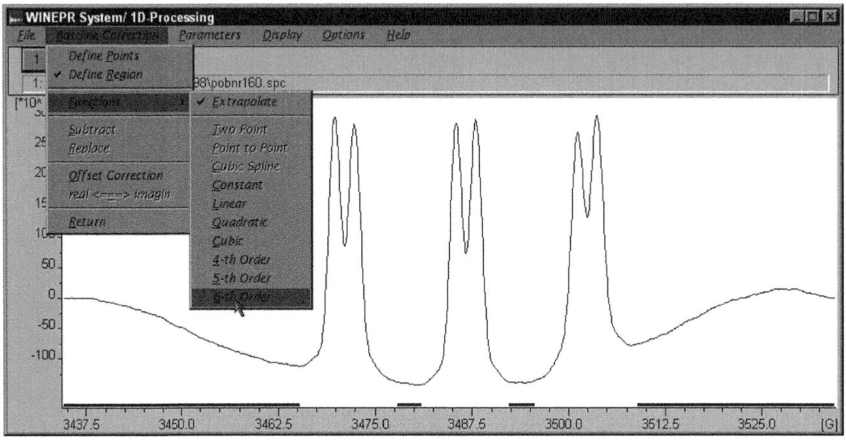

Fig. C.4 Absorption spectrum before baseline subtraction

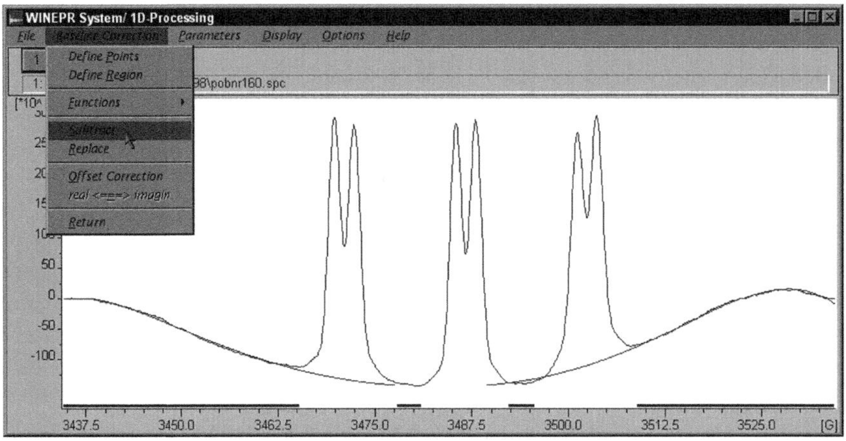

Fig. C.5 Fitting and subtracting the baseline

defined for this spectrum are marked with dark colored bars at the bottom of the window (Fig. C.4). It is important not to include any points from the actual EPR spectrum in the region that is defined for baseline correction. After defining the regions, calculate the baseline using a function from the "Functions" menu within the "Baseline Correction..." drop-down menu. In this example a sixth order polynomial was selected to calculate the baseline based on the selected regions. The fit of the baseline appears overlapping the spectrum in Fig. C.5. Next, select "Subtract" from the "Baseline Correction..." drop-down menu. The absorption spectrum with a corrected baseline then appears (Fig. C.6). At this point, if the user is satisfied with the result, click "Return" which returns to the "1D-Processing" menu. Select "Differentiate" under "1D-Processing" to get back to the first derivative representation of the EPR spectrum.

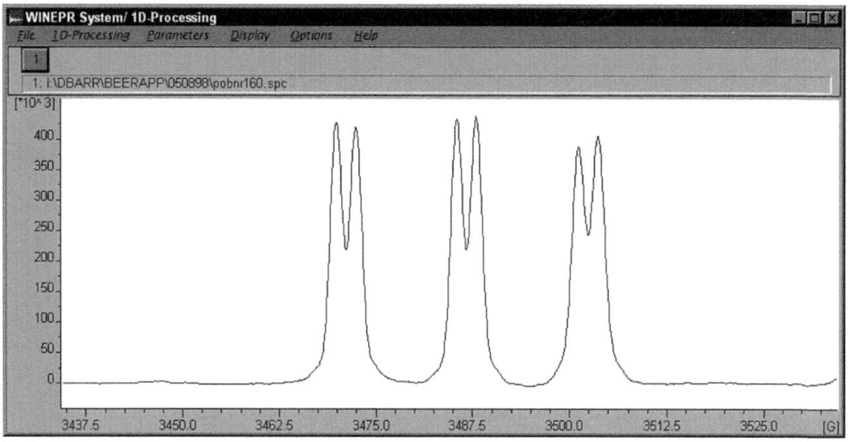

Fig. C.6 The absorption spectrum after baseline subtraction

Calculate the Double Integration

Click "Return" in the baseline correction drop-down menu. Next, select "Integrate region..." from the "1D-Processing" drop-down menu (Fig. C.7). The menu bar changes to include the "Integration" drop-down menu. Now click on "Integral Type" from the Integration drop-down menu and select "Double." Next, click on "Define Integrals." Define the region by clicking the left mouse button with the arrow pointing to the far left side of the spectrum. Then move the arrow to the outermost point on the right side of the spectrum and click the right mouse button. When the right mouse button is clicked, the double integration for the selected region appears, superimposed on the EPR spectrum (Fig. C.8).

Obtain the Double Integration Value

Click the "Report..." selection in the integration drop-down menu to see the numerical value for the double integration (Fig. C.9).

The DI/N value is the double integrated intensity of the EPR spectrum that has been normalized to account for the conversion time, receiver gain, number of data points and sweep width. The report can be printed directly or saved as an ASCII file.

This example demonstrates one possible solution for improving a double integration from a spectrum that had either a slight baseline drift, or a very broad background signal. When spectra are recorded for quantitative studies, extra care should be taken when the spectra are acquired, to minimize baseline drifts and background signals. Make sure the cavity is clean and free of background signals.

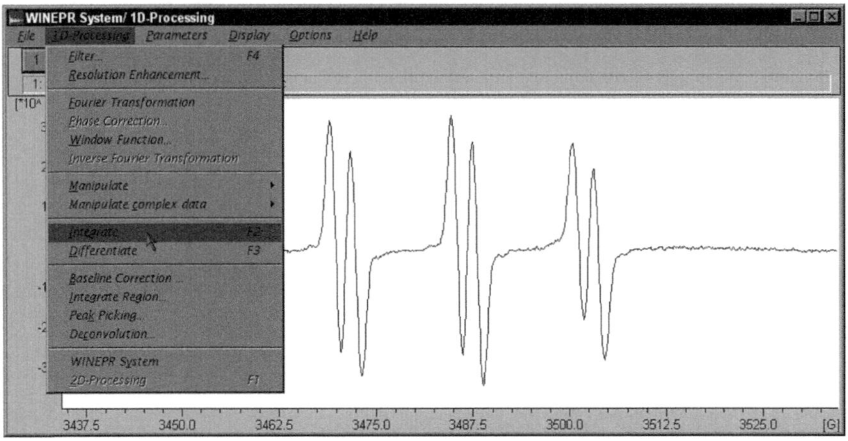

Fig. C.7 Selecting the "Integrate Region..." command

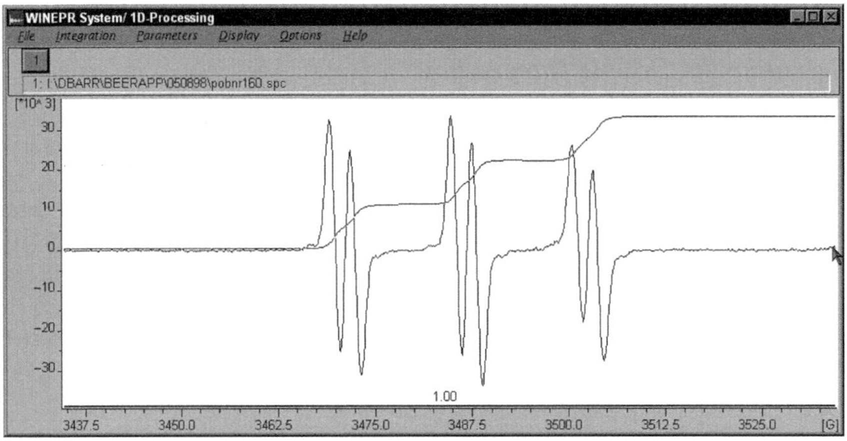

Fig. C.8 The double integration appears superimposed on the first derivative EPR spectrum

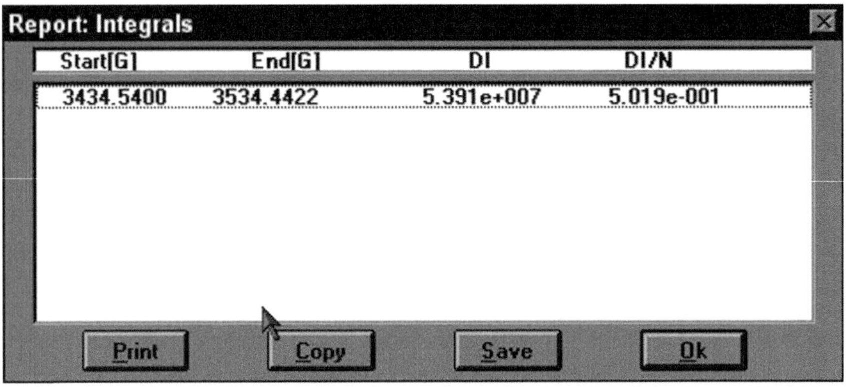

Fig. C.9 The "Report" window shows the numerical value for the double integration

For long acquisitions, use signal averaging to minimize baseline drift. If very low signal to noise is unavoidable, or if there is an overlapping signal that is known to not be part of the spectrum for which the integral is desired, perform the double integrations on simulated spectra.

Use of a Simulation to Improve Peak Intensity Measurements from Noisy Spectra

Sometimes no matter how hard a user tries to optimize the acquisition parameters, the only spectra obtained are very noisy because the signal is very weak. Even with advances in the sensitivity of EPR spectrometers, this remains a problem because these new highly sensitive spectrometers have allowed researchers to measure samples with lower and lower spin concentration. The following provides an example where a simple peak "fit-and-replace" routine was able to greatly improve the repeatability of the results from a peak intensity measurement.

In the following example the peak and trough regions of a noisy spectrum were fitted with a third order polynomial. These portions of the spectrum were replaced by the fit data and then a peak pick routine was employed to determine the EPR intensity of the spectrum:

1. Acquire the data after optimizing for the highest signal-to-noise possible (Fig. C.10).

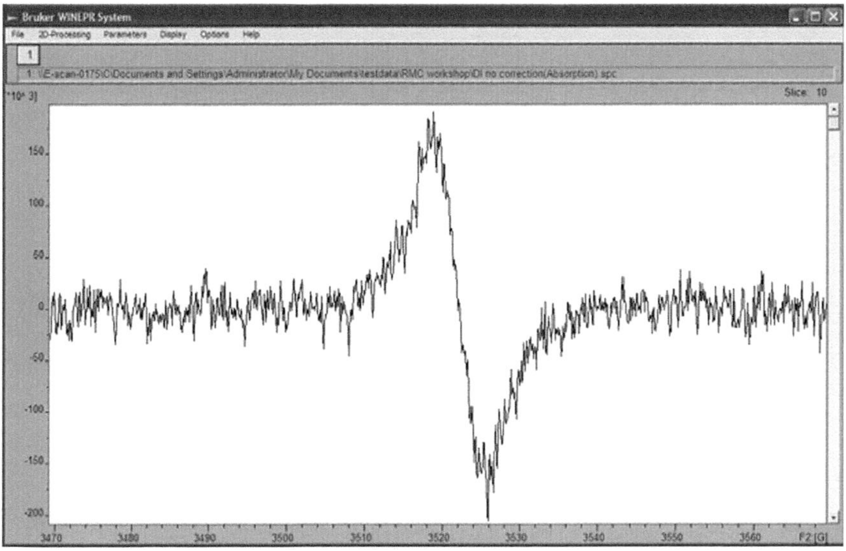

Fig. C.10 The original (relatively noisy) spectrum

2. Define the peak and trough regions and fit them with a polynomial (Fig. C.11).
3. Replace the peak and trough regions with the fitted data and perform the peak pick to obtain the intensities (Fig. C.12).

The above sample was measured ten times. The peak intensity data are shown in Fig. C.13. The fit-and-replace routine reduced the RSD (Relative Standard Deviation) for the measurement (Fig. C.13) by more than a factor of two!

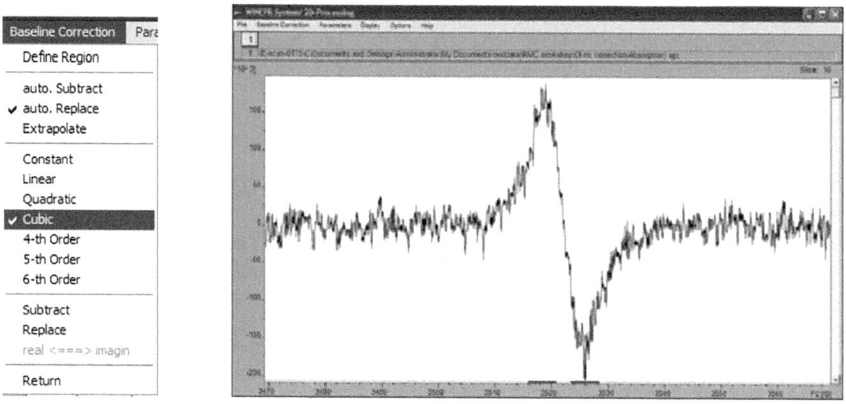

Fig. C.11 The peak and trough regions are fitted to a polynomial

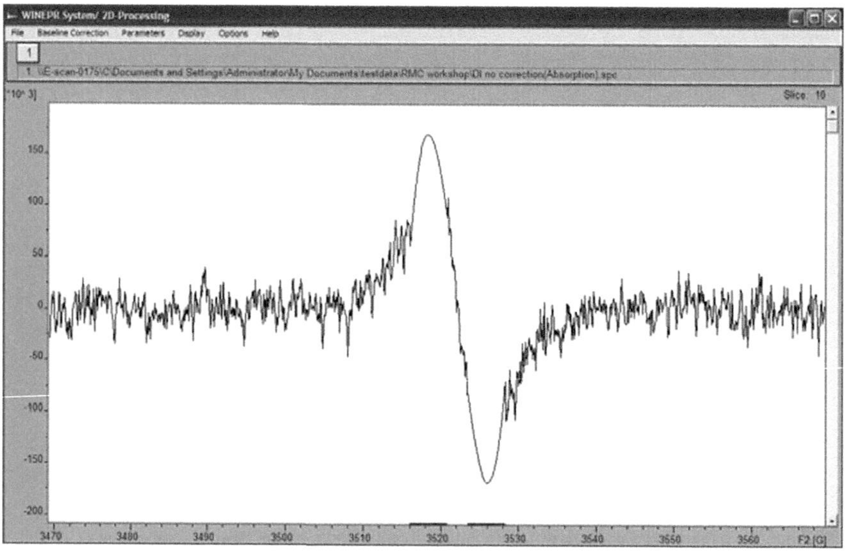

Fig. C.12 The result spectrum contains the new peak and trough fitted data

4					
5	Original Peaks	Intensity:		Smoothed Peaks	Intensity:
6	1	331445		1	330117.1
7	2	330656		2	332698.5
8	3	352362.7		3	331539.6
9	4	358016		4	347550.6
10	5	324885.3		5	338166.2
11	6	339765.3		6	338820.9
12	7	358592		7	336216.9
13	8	348058.7		8	337959.1
14	9	350154.7		9	334959.8
15	10	363301.3		10	338094.8
16					
17	**Average**	**345723.7**		**Average**	**336612.3**
18	**StdDev**	**13314.5**		**StdDev**	**4911.2**
19	**RSD**	**3.85%**		**RSD**	**1.46%**
20					

Fig. C.13 Peak intensity comparison for peak intensity data from the original data spectra compared with that obtained after the fit-and-replace routine

Additional Techniques to Improve Double Integration Results

There are a number of additional ways that one can improve the quality of spectra for making either peak intensity or double integral measurements. Below is a list of some of these methods:

1. Simulation
2. Moving average filter
3. Fit-and-replace
4. FFT combined with window function filter
5. Least squares fitting
6. Cross correlation

Appendix D: Quantitation of Organic Radicals Using Tempol

The purpose of this appendix is to describe a general method for quantitating organic radicals such as a spin-trap adduct in nonviscous solvents using the stable nitroxide 4-hydroxy tempo (Tempol) as a spin standard. There are multiple sample-related and instrument-related factors that must be considered in designing EPR quantitation experiments. The details of these factors and many of the precautions one must be aware of are discussed in the main body of this book. Probably the biggest challenge in quantitative EPR is finding a spin standard that has similar EPR behavior to that of the sample. Experiments with nitroxides have an advantage, in that, standards such as Tempol are chemically very similar, and thus, they exhibit similar EPR behavior. In addition, Tempol is stable, soluble in a variety of solvents, and is commercially available. This appendix is a sample experimental procedure in which a quantitative EPR comparison was made between the spin trapped DMPO/hydroxyl radical adduct (DMPO/OH) and a Tempol solution of known concentration.

Determine the Concentration of the Tempol Solution

Before doing any EPR, the concentration of the Tempol standard should be carefully determined. Prepare a nominal 100 mM solution of Tempol in water by weighing the appropriate amount of solid using an analytical balance. The concentration should be verified using the optical absorption of the Tempol. Morrisett (1976) reported an extinction coefficient of 1,440 M^{-1} cm^{-1} at 240 nm for Tempol in ethanol. Kooser et al. (1992) describe a titration method using ascorbate that allowed them to determine the extinction coefficient of Tempol in water at 429 nm (13.4 M^{-1} cm^{-1}). While many impurities may absorb at 240 nm, the absorbance of a Tempol solution at 429 nm is very likely to be specific for the nitroxide moiety. Therefore, the method described by Kooser et al. is recommended to determine the concentration of the Tempol standard. The extinction coefficient of 13.4 M^{-1} cm^{-1} gives good agreement with gravimetric determinations corrected for the purity of Tempol purchased from Aldrich Chemical Co., Milwaukee, WI, Aldrich No. (17,614-1). If the concentration calculated with this value of ε does not match the concentration based on the gravimetric preparation, the ascorbate titration described by Kooser et al. should be performed.

Prepare Several Dilutions of the Stock Tempol Solution

Starting with the nominally 100 mM stock solution, make two 1:10 dilutions. Use the second dilution (nominally 1 mM) to make dilutions to 300, 200, 100, 75, 50, 25, 12.5 and 5 μM.

Record the EPR Spectra of the Tempol Dilutions

The following parameters were selected based on the optimization of settings for Tempol in Appendix A. Most importantly, the microwave power is significantly below the saturation level in the case of a Bruker ER 4119HS cavity. If a different cavity is used, an appropriate microwave power should be determined by following the optimization procedure outlined in Appendix A.

The spectrometer settings were as follows:

Microwave frequency	9.78 GHz
Modulation frequency	100 kHz
Microwave power	5 mW
Modulation amplitude	1 G
Time constant	2.56 ms
Scan time	2.62 s
Conversion time	2.56 ms
Number of scans	64
Field sweep	100 G
Center field	3,481
Receiver gain	2×10^3
Number of data points	1,024

Determine the Double Integrals of the EPR Spectra from Each of the Dilutions

This step is very important. To obtain meaningful quantitative results, the double integrals must accurately represent the EPR absorption of the samples. Baseline drift, background signals, and a low signal to noise ratio will all decrease the accuracy and reproducibility of the double integrations. For the purpose of demonstration, the signals in this appendix were very strong which made the double integration quite accurate (see Fig. D.1).

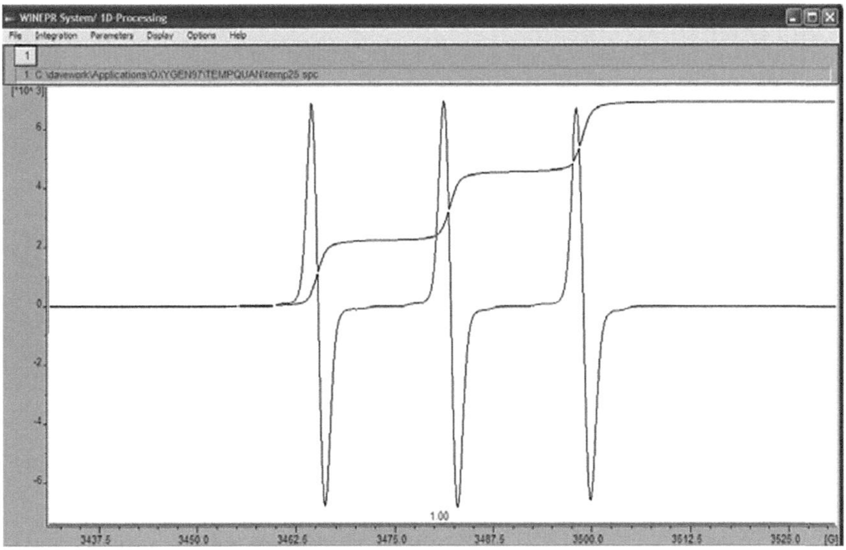

Fig. D.1 Double integral from one of the Tempol concentrations used in the standard curve

Make a Standard Curve of Double Integrated Intensity Versus Tempol Concentration

The standard curve (see Fig. D.2) should be made using as many replicates at each Tempol concentration as is necessary to give reasonable statistical significance. It is helpful to prepare the 100 mM stock solution several times and repeat the dilution process. This will give an idea of the error contributed by weighing, measuring the 429 nm absorbance, diluting the stock solution, etc. The double integrated intensity of the EPR spectra should increase linearly as a function of the Tempol concentration. A linear regression analysis can be used to get an R^2 (i.e., the square of the correlation coefficient between the observed and predicted Tempol values) and slope for the curve.

Prepare a DMPO/OH Sample

The next step is to prepare the radical adduct sample. For this example a simple Fenton reagent (i.e., ferrous ammonium sulfate and hydrogen peroxide) was used. The EPR spectrum was recorded using the settings described in Step 2 because previous experiments had shown that the linewidths and power saturation were similar to that for Tempol.

Use WIN EPR to Determine the Double Integral of the DMPO/OH Spectrum

The double integrals for the DMPO/OH sample should be calculated using the same procedures as for Tempol (Fig. D.3). Next, locate the position of the double integral

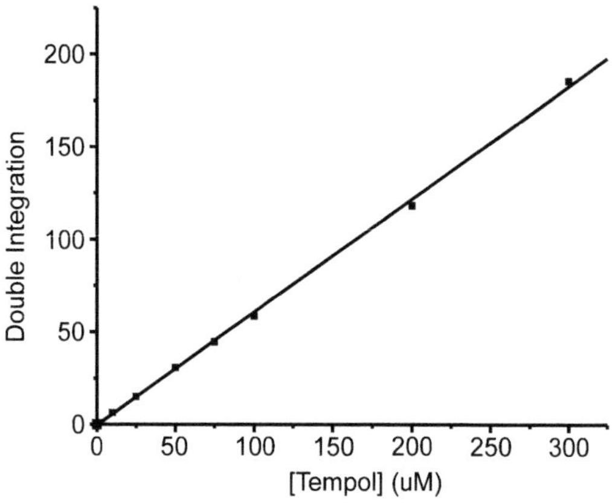

Fig. D.2 Typical standard curve for the double integrated intensity of Tempol

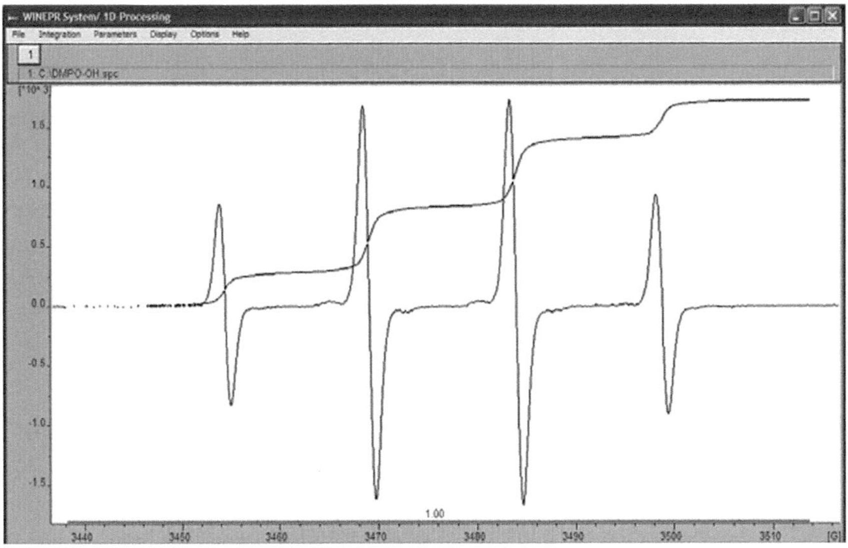

Fig. D.3 Double integration of experimental spectrum

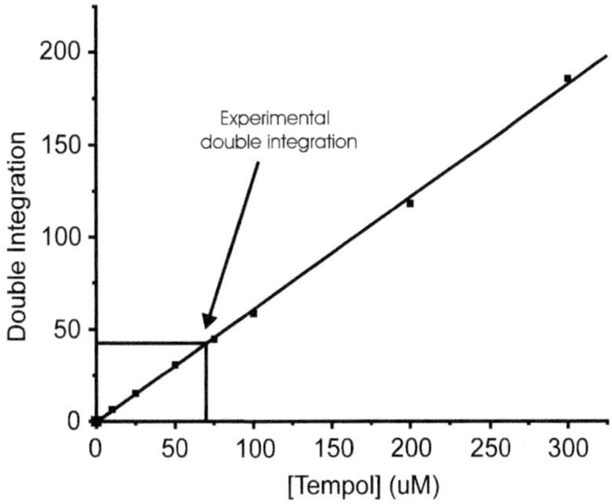

Fig. D.4 Locate the position of the double integral of the DMPO/OH sample on the Tempol standard curve

for the DMPO/OH sample on the standard curve for Tempol (Fig. D.4) Alternatively, the slope from linear regression analysis can be used to calculate the concentration of DMPO/OH that was present in the reaction.

Summary

This appendix presents a simple method for quantitating nitroxide radical adducts. The general method has been used by many researchers in the field of biological EPR. Experiments with nitroxides provide a situation where the spectroscopist can be confident that the EPR behavior of the sample (i.e., spin trap adduct or spin label) and the standard are very similar. With other types of samples it may be difficult to find a standard that has sufficiently similar EPR behavior. Therefore, it is important to read Chap. 10 in this book. If the sample to be quantitated is not a freely tumbling nitroxide, be certain that an appropriate standard is selected.

Appendix E: Using a Reference Standard for Relative Intensity Measurements

In recent years there has been growing use of EPR as a simple detection device for various processes. In these situations the researchers or technologists have found that EPR provides a unique alternative because it is specific, sensitive and in many cases provides the most rapid and least labor intensive method. Some examples of industries employing EPR for routine measurement include: the pharmaceutical, radiation processing, brewing, food and polymer industries (just to name a few). In each of these cases the most important factor in the end is the reproducibility of the measurement. There is a need to have confidence in the measurement from day-to-day, and to be able to get a reasonable estimate of measurement uncertainty (hopefully low).

The use of a stable reference standard is often the best way to provide the reliable day-to-day measurement consistency that these applications all require. The system is configured so that the EPR spectrum of the unknown samples can be measured simultaneously with the reference sample. Often the EPR intensity ratio of the unknown sample to that of the reference sample is used rather than the intensity of the sample itself. This appendix provides an overview of procedures where a stable EPR reference standard is employed.

Why use an EPR Reference Standard?

In this book many of the effects of instrumental settings and sample environment on the EPR signal have been discussed. Variations in Q-value of the cavity, detector variations, etc. can all compromise the ability to make consistent measurements (even on the same sample). By using a permanently mounted (or at least a reproducibly mounted) EPR reference standard, the impact of many of these unavoidable instrumental variances on the measurement can be minimized. This is

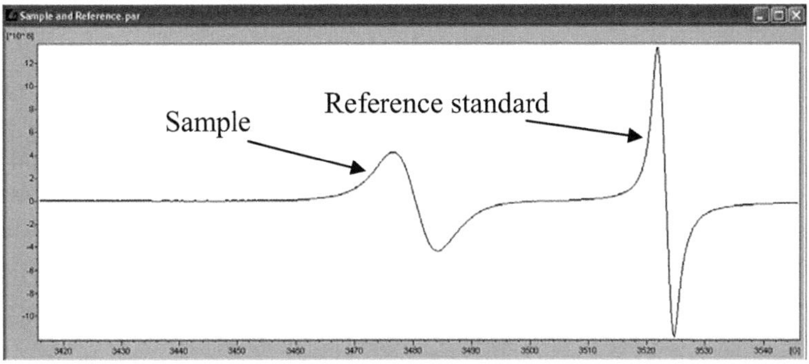

Fig. E.1 Example of an EPR reference standard that has a resonance position that is ~40 G upwards in field from a g = 2 signal

done by making an EPR intensity measurement on both the sample and the reference at the same time (Fig. E.1), and then using the ratio as the primary result.

$$\text{Measured Value} = \frac{\text{Sample Intensity}}{\text{Reference Standard Intensity}}$$

When this is done, small deviations in the sensitivity of the instrument due to temperature, humidity, electrical variances, etc are not introduced into error of the measurement. The reference standard experiences the same changes as the sample. So, although the absolute intensity values may vary, the ratio of sample intensity to reference intensity remains the same.

Properties of an Ideal EPR Reference Standard

1. *Stability* – The reference standard material must be stable to allow reproducible measurements over extended periods of time. Stability is defined here as: chemical stability, photo stability, physical stability, and of course unvarying intensity
2. *Isotropy* – The position of the reference standard's EPR spectrum must not vary as a function of its orientation in the magnetic field.
3. *Field for resonance* – The reference standard's EPR spectrum must not substantially overlap the measured sample's EPR spectrum. Ideally both spectra should be acquired in the same field scan.
4. *Cost* – The reference standard material should be low in cost and easy to prepare reproducibly.
5. *Ease of use* – The reference standard material should be easy to configure in various holders or other devices that allow reproducible positioning in the microwave cavity.

Positioning of the Reference Standard is Critical

For EPR to be successful as a routine instrument in an industrial QA/QC (Quality Assurance/Quality Control) environment the measuring set-up needs to be simple and provide very reproducible measurements from day-to-day. Once a good reference standard material is selected, the next goal is to encase the standard in a sample holder or part of the cavity that will assure that its position does not change. Below are some examples of reference standard holders that have been successful in providing very reproducible measurements (Fig. E.2).

Testing the Measurement Reproducibility of an EPR Reference Standard

A proposed reference standard setup should first be tested alone, by making repeatability and reproducibility measurements. Repeatability measurements are

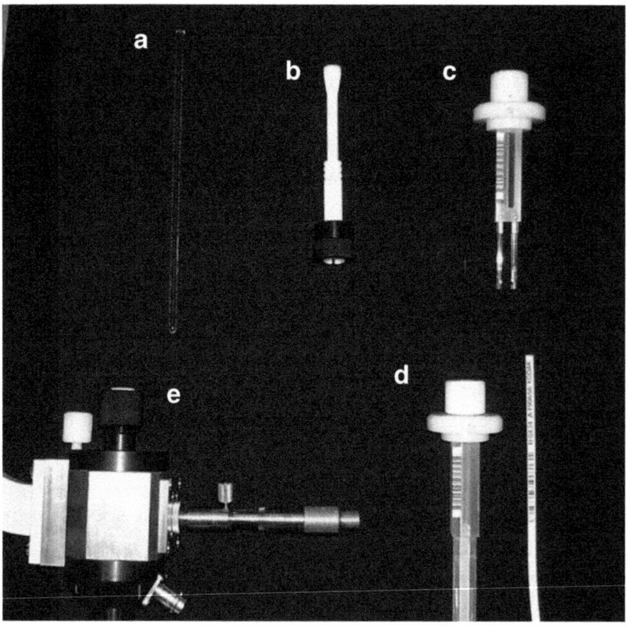

Fig. E.2 Examples of devices for holding an EPR reference standard. (**a**) is a simple capillary holder with a groove that can be packed with powdered reference standard material. (**b**) is a reference standard pedestal that can be fixed into the cavity in three different positions. (**c**) is a reference standard containing a tube holder for the e-scan benchtop spectrometer. (**d**) is an alanine film dosimeter holder. (**e**) is a cavity mounted micrometer device that permits moving a reference standard capillary in and out

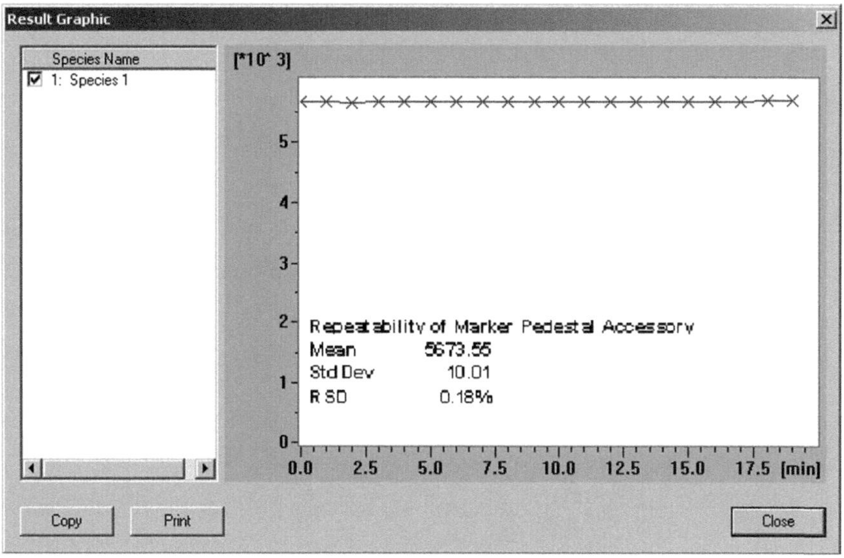

Fig. E.3 Peak intensity data for a pedestal mounted EPR reference standard

defined here as consecutive measurements made without moving the reference standard out of the cavity. Reproducibility is defined as consecutive measurements made where some sort of movement or repositioning of the reference standard is made in between measurements (such as might occur in the routine assay). Figures E.3–E.5 show examples of a repeatability and reproducibility measurements performed with the Bruker reference standard pedestal accessory.

Repeatability Test

In this test the reference standard containing pedestal was locked into one of the 4 mm positioning grooves and 20 consecutive field scans were performed (35 s delay). The pedestal was left in place throughout the 20 scans. This test can also be done for longer periods (such as overnight or over several days).

Reproducibility Test

In this test the EPR spectrometer was configured to perform 20 field scans with a 35 s delay between scans. After every other field scan (during the 35 s delay), the reference standard containing pedestal was removed completely from the cavity,

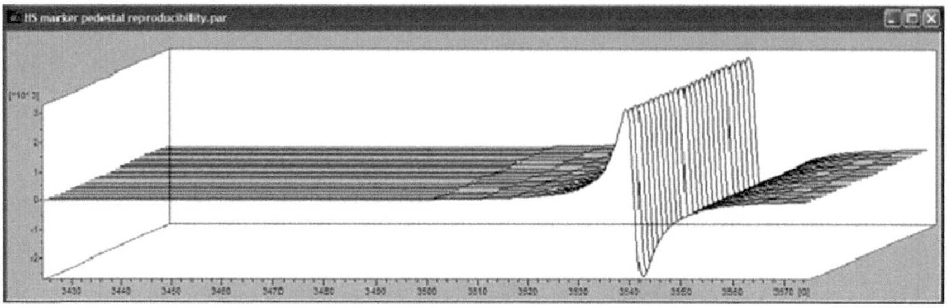

Fig. E.4 Reproducibility test for the reference standard pedestal

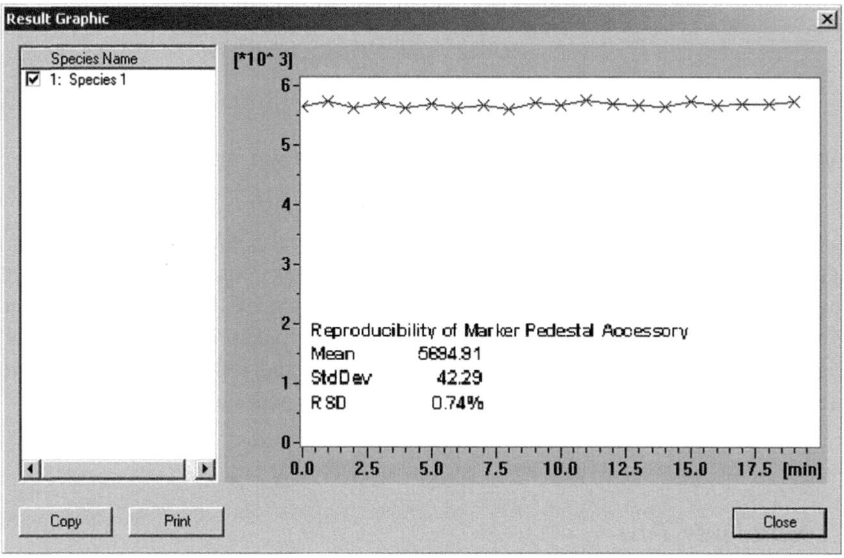

Fig. E.5 Peak intensity data for a pedestal mounted EPR reference standard obtained in a reproducibility test

and re-inserted locking it each time into one of the three 4-mm positioning grooves. The results of the test are summarized in Fig. E.5.

Summary

EPR reference standards have proven invaluable for making routine EPR measurements in industrial processes. For example the Bruker e-scan alanine dosimetry system is capable of measuring irradiation dose from a few Gray to 100 kGy with a total measurement uncertainty of ±3%. The instrument's measurement reproducibility, itself, is less than ±1%. Such precision and low uncertainty, provide substantial annual savings for radiation processing facilities and make EPR a viable method for their QC/QA work.

Appendix F: Example Procedure for Measuring Signal-to-Noise Ratio

Signal-to-Noise Testing for Spectrometer Maintenance

The signal-to-noise ratio (S/N) test is an important part of routine spectrometer maintenance. It is also helpful in diagnosing possible problems that may be encountered especially when dealing with very weak signals or signals to be quantitated. For many years spectrometer manufacturers have tested S/N with specified power, modulation amplitude, and other parameters using a particular vendor-supplied "weak pitch" sample. Some historical background is provided in Eaton and Eaton (1992). This appendix is specific to Bruker spectrometers with an ER 4119HS standard cavity and the weak pitch sample shipped with the spectrometer. The test measures the EPR signal intensity (peak-to-peak height) of the weak pitch sample at low microwave power (12 dB) and then measures the noise level under the same conditions except higher microwave power (0 dB) and higher receiver gain to characterize the noise better. The formula for calculation of signal-to-noise ratio (S/N) is:

$$\frac{S}{N} = \frac{A_S}{A_N} \times \frac{G_N}{G_S} \times \sqrt{\frac{P_N}{P_S}} \times \frac{2.5}{\sqrt{T} \times C} \tag{F.1}$$

where A_S and A_N are the peak to peak height of the weak pitch and amplitude of the noise respectively. G_S and G_N are the receiver gains used in signal and noise measurements respectively – their ratio is used to correct for the different gain settings use in the signal and noise data acquisition. P_S and P_N are the powers used in the two measurements: the square root of the ratio of powers is used to correct for the power difference. The factor of 2.5 translates the peak-to-peak noise level to a RMS (Root Mean Square) noise level. T is the time constant (in seconds): the square root of the time constant is used to normalize the S/N to a 1 s time constant. C is the weak pitch correction factor that is printed on the label of the Bruker weak pitch sample. The standard instrument settings for signal and noise measurements are listed in Table F.1. Note that the weak pitch line is severely over-modulated in this test to blur out sample-to-sample differences in line width and line shape. In Xepr there is a built-in subroutine to measure the signal to noise ratio which has the default values of standard settings. To measure the amplitudes of the signal and noise on a print out

Table F.1 Instrument setting for signal/noise measurement

Parameter	Signal	Noise
Modulation amplitude (G)	6.0	6.0
Modulation frequency (kHz)	100	100
Receiver gain	2.0×10^4	5.0×10^4
Phase	0	0
Time constant (ms)	1,310.72	1,310.72
Conversion time (ms)	163.84	163.84
Center field (G)	3,480	1,500
X-axis setting	Field sweep	Time scan
Sweep width (G)	50	–
X resolution (points)	1,024	1,024
Microwave attenuation (dB)	23	0

by hand, make sure that the same scale is used for both signal and noise spectra. Otherwise the result should be multiplied by the ratio of the plot scales.

Spectrometer Settings for Signal/Noise Measurements Using the Bruker ER 4119HS Cavity

Measuring the Signal to Noise Ratio

Step 1. Open the Signal/Noise Ratio Test window under the Acquisition drop-down menu (see Fig. F.1). The window has two empty spectra and each one contains a set of default parameters for signal or noise measurement. Click either one of the windows with the left mouse button to activate that window. The parameters shown on the right will be assigned to that measurement.

Step 2. Enter the calibration factor printed on the label of the weak pitch sample into the Weak Pitch factor box (see Fig. F.2).

Step 3. Activate the signal measurement. Click the signal window (the upper one). A blue bar will appear on the right upper corner. Check the parameter settings by opening the Standard Parameter dialog box. The parameters should look like those in Fig. F.3.

Step 4. Set a time delay before each field sweep. Since a very long time constant is used, set a delay time of 2–5 s to avoid overshoots or undershoots in the first few data points when the spectrum is acquired. Open the Experimental Options dialog box (found in the Parameter drop-down menu) and set the Delay before each sweep option, and enter a delay of 2–5 s (see Fig. F.4). It is useful to select the MW Fine Tune before each sweep option to ensure the acquisition is made under proper coupling conditions.

Step 5. Acquire a signal spectrum. Click the RUN button in the tool bar to acquire a weak pitch spectrum (see Fig. F.5). If the spectrum is off center, use the

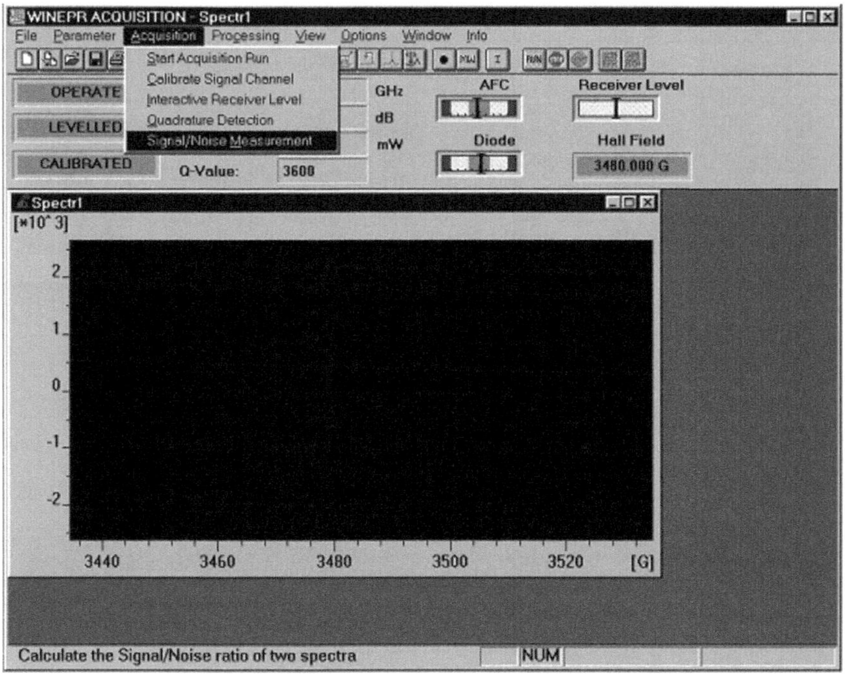

Fig. F.1 Open signal/noise measurement window

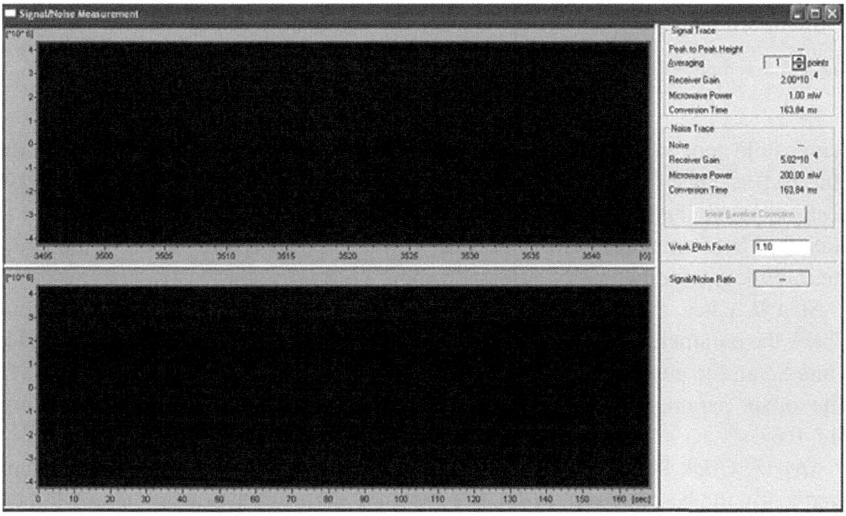

Fig. F.2 Enter the weak pitch factor

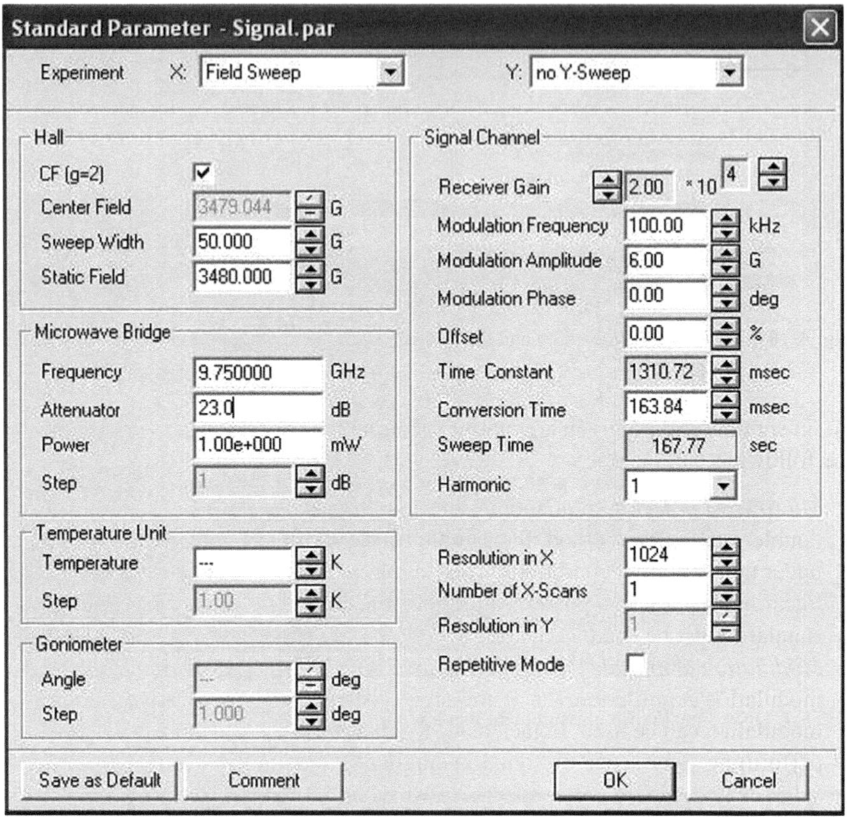

Fig. F.3 Example parameters for signal measurement (Bruker ER 4119HS cavity)

center field tool to set the correct field center. If there is a large offset, open the "Interactive Spectrometer Control" dialog box and adjust the offset to the proper position where the indicator of the Receiver Level is in the middle. Do not forget to click the "Set Parameters to the Spectrum" button and move the pointer to the signal measurement window and click the left mouse button again.

Step 6. Click the lower window to activate the noise measurement window. Check the parameters. Open the parameter dialog box. Make sure the X axis is set to Time Scan, the power is 200 mW, gain is 5×10^4 and the field center at 1,500 G. The other parameters should be similar to those in signal measurement (see Fig. F.6).

Step 7. Click the RUN button in the tool bar to acquire the noise spectrum. Frequently the baseline will drift since 200 mW microwave power is going to heat up the cavity and the sample. Wait a few minutes to achieve thermal equilibrium. Check the tuning and coupling of the system. Retune the system if necessary. There may be a rather large baseline offset due to the excessive power and high gain.

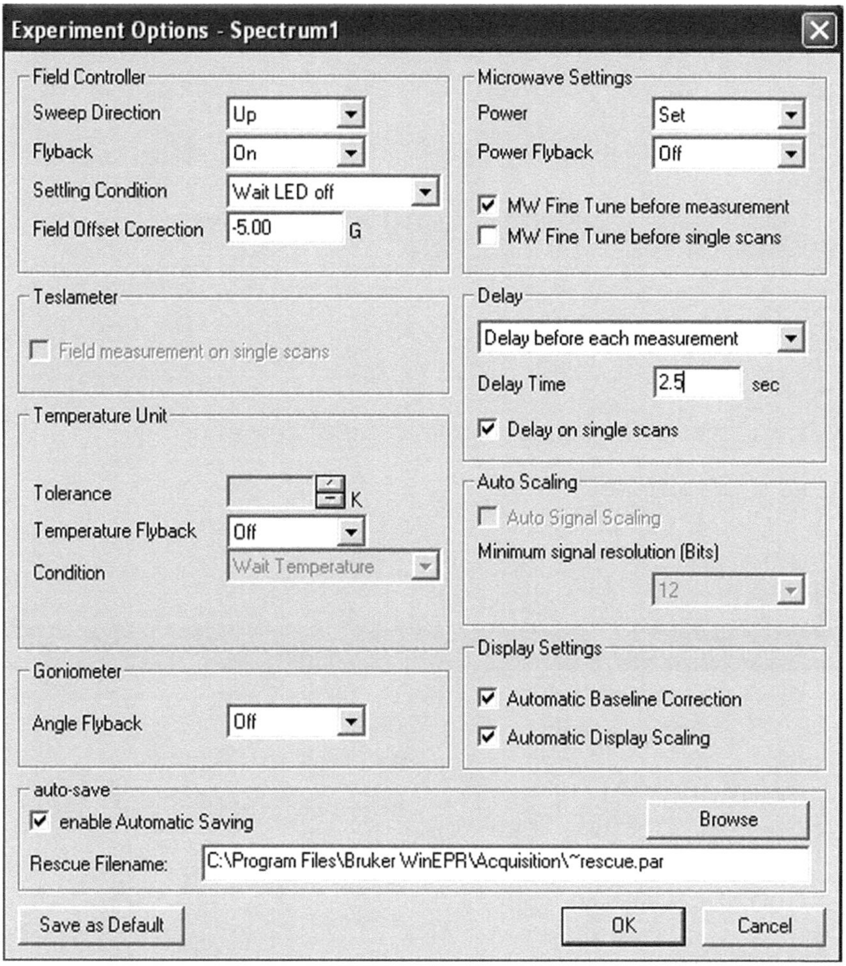

Fig. F.4 Set experimental options

Use the interactive box to make the offset adjustment so that the indicator of the receiver level is in the middle. Click the left mouse button on Set Parameters to Spectrum, move the pointer to the noise measurement window, and click again. If the signal overshoots or undershoots, a 2–5 s delay time can be set in the Experimental Options box as in Step 4.

Step 8. Acquire a noise spectrum. Click the RUN button in the tool bar and acquire the noise spectrum again. Two horizontal lines will automatically emerge indicating the noise level. If the baseline still drifts, the linear baseline correction button can be used to compensate for linear drifts.

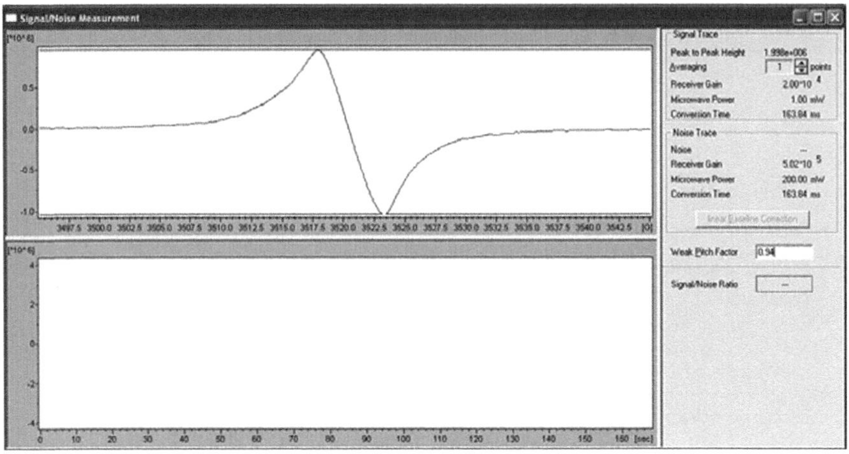

Fig. F.5 Signal measurement

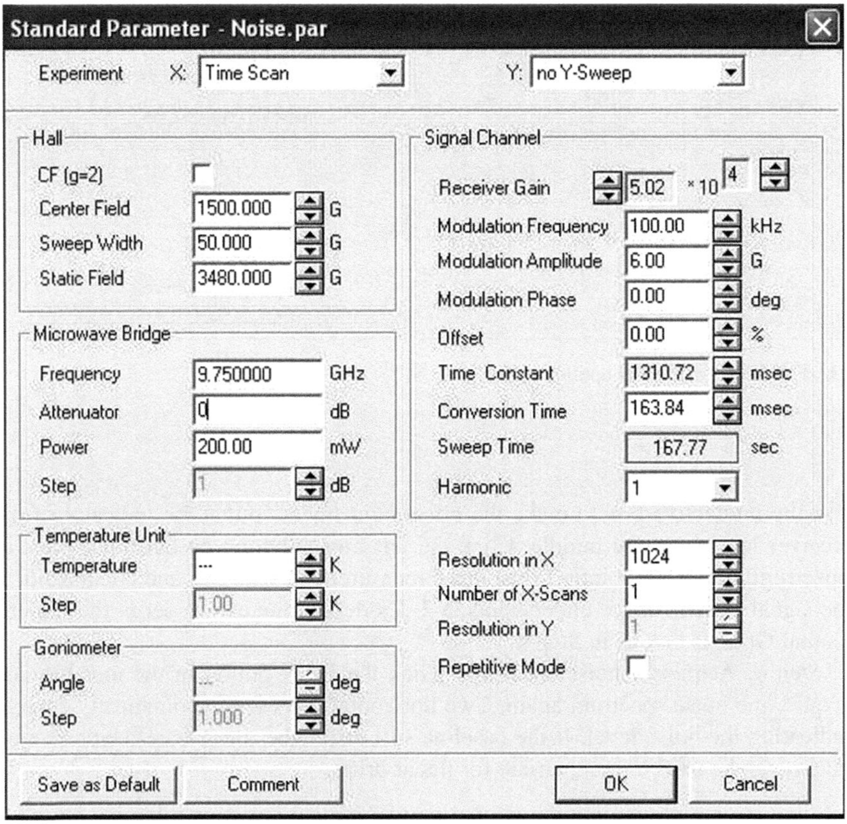

Fig. F.6 Parameters for noise measurement

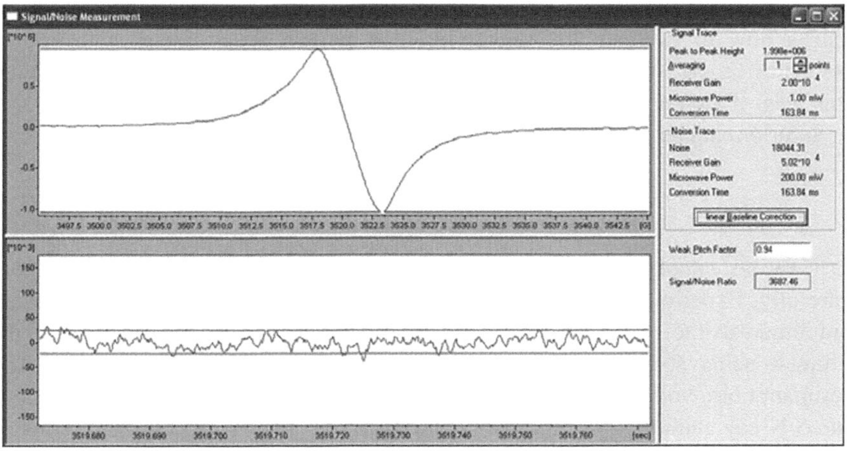

Fig. F.7 Noise measurement and the final result

Step 9. Check the S/N ratio. On the right panel the results of the signal intensity and noise level measurements appear automatically. At the bottom of the panel, the automatically calculated signal to noise ratio will be displayed in the box (see Fig. F.7). The signal to noise ratio should meet the specification indicated for a particular instrument and cavity (check the instrument's documentation for the S/N specification). If there is a change in the S/N for the instrument, a service representative should be consulted.

Appendix G: How Good Can It Get: Absolute EPR Signal Intensity

Throughout this discussion, it has been inherent that the better the S/N, the better the quantitative accuracy possible. The question, then, is what S/N can EPR aspire to? The first quantitative EPR performance criterion was a weak pitch S/N = 20 on the Varian V4502. EPR has come a long way since then. The S/N for weak pitch under comparable conditions in a TE_{102} cavity has improved approximately linearly with time (Eaton and Eaton 1992).

Why did this book not start with a measurement of the absolute EPR signal intensity, instead of leaving it to the last appendix? Hyde's 1962 statement (as reported by Alger 1968, p. 200) that "of all the measurements one can make with EPR equipment, the determination of absolute spin concentration is the most difficult" remains true more than 47 years later. This section presents the background needed to estimate the ultimate sensitivity of spectrometers, and cites papers that illustrate the state of the art in absolute concentration measurements. It shows how to estimate the sensitivity of a perfect spectrometer with known resonator Q, etc., and how real losses and active components prevent achieving this ideal (Eaton et al. 1998; Rinard et al. 1999a, b, c, 2002a, b, 2004).

In prior chapters the EPR signal has been expressed in the form:

$$V_S = \chi'' \eta Q \sqrt{PZ_0} \tag{G.1}$$

Now, to calculate the S/N, the terms χ'', η, and P have to be defined more carefully. Experimentally, the S/N measurement is a ratio of the maximum signal amplitude to the rms noise. In the calculations that follow, the estimate of the ultimate achievable S/N includes a calculation of the thermal noise power. The resultant noise voltage is an rms value, which is appropriate for the denominator of the S/N calculation. Note that the equation for V_s also is expressed in terms of microwave power, so this would yield an rms value for B_1. As discussed in Chap. 8, the filling factor calculation uses the linearly polarized B_1, but only the circularly polarized component creates the EPR signal. As explained in more detail below, to reduce confusion, the factor of 1/2 is included explicitly in the calculation of V_s, rather than in the calculation of η. The third term to discuss is the spin susceptibility. Ideally, one would use the full line shape function, and the fraction that is detected in the CW measurement, in the calculation. In this chapter, an approximation is introduced by using the expressions from Weil and Bolton (2007) for line shapes (539ff). It is assumed that the line shape is Gaussian and that the modulation amplitude will be approximately equal to the line width, so that the entire sample magnetization will be measured. Further, it is assumed that there is no power saturation of the spin system.

The Spin Magnetization, M, for an Arbitrary Spin, S: Definitions

The spin magnetization is

$$\mathrm{M_0 = H_0 \chi_0 = \frac{B_0}{\mu_0} \chi_0.} \tag{G.2}$$

$$\mathrm{M_0 = N_0 \frac{\gamma^2 \hbar^2 B_0 S(S+1)}{3k_B T} \ JT^{-1} m^{-3} \left(= Am^{-1}\right)} \tag{G.3}$$

So, M/H is unitless, as required.

For S = 1/2,

$$M_0 = N_0 \frac{\gamma^2 \hbar^2 B_0}{4k_B T} = N_0 \frac{g^2 \beta^2 B_0}{4k_B T_{sample}} \tag{G.4}$$

Definitions and units in these equations:

$$g\beta = \gamma\hbar \tag{G.5}$$

$\gamma = 1.7608 \times 10^7$ rad s^{-1} G^{-1} $= 1.7608 \times 10^{11}$ s^{-1} T^{-1}, $\hbar = 1.0546 \times 10^{-34}$ joule s rad^{-1}, N_0 is the number of spins per unit volume. (In some publications, the number of spins per unit volume is split into two terms, N, the number of spins in volume V, and V is the volume of sample in m^3.) The static magnetic field $B_0 = \omega_0/\gamma$. S is the electron spin, which is 1/2 in the calculations in this appendix. $k_B = 1.3806 \times 10^{-23}$ joule K^{-1} is Boltzmann's constant. T is the temperature of the sample in K. The permeability of vacuum, $\mu_0 = 4\pi \times 10^{-7}$ $T^2J^{-1}m^3$.

The magnetic susceptibility of the sample, χ'' (dimensionless), is the imaginary component of the effective RF susceptibility. It is assumed that the line is Gaussian, which is a reasonable approximation for a nitroxide or weak pitch sample, since there are unresolved hyperfine interactions. The assumption makes the calculation simpler than using a more realistic line shape function. For a Lorentzian line, with width at half height = $\Delta\omega$ at resonance frequency, ω,

$$\chi'' = \chi_0 \frac{\omega}{\Delta\,\omega}, \quad \text{where } \Delta\omega \text{ is the line width.} \tag{G.6}$$

The formula in (G.6) requires use of the absorption line width, so the peak-to-peak derivative width is multiplied by $\sqrt{3}$. Although nitroxide and weak pitch lines are not Lorentzian, this simple approximation provides a reasonable estimate of χ'', but small adjustments can be made for Gaussian or mixed line shapes.

At X-band, for ca. 9.4 GHz, B_0 = ca. 3,360 G = 0.336 T. To get units right in these calculations use the relation (from Wertz and Bolton, 1972) that the units of gauss are $\frac{erg}{Gcm^3}$, and convert to SI units. Then, use the susceptibility formula in the following form:

$$\chi'' = \chi_0 \frac{\omega}{\Delta\omega} = \frac{N_0\gamma^2\hbar^2\mu_0}{4k_BT}\frac{\omega}{\Delta\omega} \tag{G.7}$$

Units for this equation cancel

$$\frac{\text{rad s}^{-1}}{\text{G rad s}^{-1}\text{G}^{-1}}\;\frac{m^{-3}(s^{-1}T^{-1})^2(Js)^2(T^2J^{-1}m^3)}{JK^{-1}K} = unitless$$

It is convenient to combine all terms that are not specific to a particular sample to simplify multiple calculations.

$$\chi'' = \frac{N_0\omega}{\Delta\omega}\frac{(1.76 \times 10^{11})^2(1.0546 \times 10^{-34})^2(4\pi \times 10^{-7})}{4(1.38 \times 10^{-23})(295)} = 2.66 \times 10^{-32}\frac{N_0\omega}{\Delta\omega} \tag{G.8}$$

N has units of spins per cubic meter, and frequency and line width are both either Hz or G. For this example room temperature is assumed. If the line width does not change and the signal does not saturate, the sensitivity decreases inversely proportional to the temperature.

Next, calculate the signal voltage, V_S, for a "perfect" spectrometer, which does not add noise in the detection system using,

$$V_S = (1/2)\chi'' \eta Q_L \sqrt{Z_0 P} \tag{G.9}$$

where V_S is the CW EPR signal voltage at the end of the transmission line connected to the resonator, η (dimensionless) is the resonator filling factor, Q (dimensionless) is the loaded quality factor of the resonator, sometimes denoted Q_L, Z_0 is the characteristic impedance of the transmission line (in ohms, usually 50), and P is the microwave power (in W) to the resonator produced by the external microwave source.

A 1 mM S = 1/2 radical sample has about

$$\left(1 \times 10^{-3}\text{M/L}\right)\left(10^3\text{L/m}^3\right)\left(6.022 \times 10^{23}\text{M}^{-1}\right) = 6.022 \times 10^{23} \quad \text{spins/m}^3 \tag{G.10}$$

These are unusual units for chemical concentrations, but the spin concentration per cubic meter is needed for equations in SI units.

Assume a single Gaussian line with 1 G width. The goal of the calculation is the maximum signal amplitude (i.e, the peak of the line). The susceptibility to be used in the calculation of signal voltage V_S, for a 0.01 mM solution is thus,

$$\chi'' = \left(2.66 \times 10^{-32}\right)\frac{2\pi 9.4 \times 10^9}{1 \times \sqrt{3} \times 2\pi 2.8 \times 10^6}\left(0.01 \times 6 \times 10^{23}\right)(0.8) = 2.5 \times 10^{-7} \tag{G.11}$$

The 0.8 in this equation results from the assumption of a Gaussian line shape.

Nitroxide – Note that discussion of the sensitivity for nitroxides has to account for the fact that the intensity is spread over three lines due to the ^{14}N hyperfine, so the concentration is divided by 3 and the calculation is based on an effective concentration.

Signal Voltage

Equation (G.9) is the signal voltage for CW EPR, for a given sample of given concentration in a resonator with a given Q and a given filling factor for a given input power (Rinard et al. 1999a, c). Recall from the discussion at the beginning of this appendix that the 1/2 in (G.9) comes from the fact that only half of the total microwave magnetization is effective in causing EPR transitions. One of the circularly polarized components rotates in the opposite direction and has little effect on the spins.

To perform a sample calculation for a 0.01 mM nitroxide sample: assume as standard values $Q = 3{,}000$, $P = 1 \times 10^{-3}$ W, and $Z_0 = 50$ ohm. The filling factor can be approximated as the ratio of the volumes, but more accurate calculations for a Bruker rectangular TE_{102} cavity resonator are presented in Chap. 8 on Filling Factor. Using those results, assume that the filling factor, η, is about 6×10^{-3} (0.6%).

Substitution of these values into (G.9) gives the signal voltage at the end of the transmission line connected to the resonator for the 0.01 mM nitroxide sample as

$$\begin{aligned} V_S &= (1/3)(1/2)\left(2.5 \times 10^{-7}\right)\left(6 \times 10^{-3}\right)(3000)\sqrt{(50)\left(1 \times 10^{-3}\right)} \\ &= 1.7 \times 10^{-7} V \end{aligned} \tag{G.12}$$

This is the signal voltage for each line of the 0.01 mM nitroxide sample prior to amplification. The gain of the spectrometer can be ignored for the purpose of this S/N calculation, since signal and noise are amplified equally. This calculation is equivalent to assuming an ideal spectrometer that adds no noise.

Calculation of Noise

There are several contributions to the noise in the spectrometer. Ideally, the limiting noise would be the noise factor (NF) of the first stage amplifier amplifying thermal noise. However, there are also losses in the path from the resonator to the detector, which can be large at X-band, source phase noise, and the NF of the detection system. Calculation of the noise at the output of the resonator provides an estimate of ideal S/N.

The thermal noise power in a 50 ohm system, in dBm, is given by

$$\text{Pn} = -174 + 10\log(\text{bandwidth}). \tag{G.13}$$

Values in dBm are relative to 1 mW power.

Assume a 1 s filter time constant, which corresponds to ca 0.125 Hz effective noise bandwidth for a two-pole filter. Thus, $\text{Pn} = -174 + 10\log(0.125) = -183$ dBm. To convert from dBm to voltage, convert to watts by dividing Pn by 10 and taking the inverse log: Inverse log(dBm/10) $= 5 \times 10^{-19}$ mW $= 5 \times 10^{-22}$ W. Since this is power, the noise is calculated as rms V by multiplying by 50 ohm and taking the square root.

$$\text{V} = \sqrt{\text{WR}} = \left(50 \times 5 \times 10^{-22}\right)^{1/2} = 1.58 \times 10^{-10}\text{V} \tag{G.14}$$

Calculation of S/N for a Nitroxide Sample

This pair of calculations suggests a maximum S/N for the 0.010 mM nitroxide solution of

$$S/N = \frac{(1.7 \times 10^{-7})}{(1.58 \times 10^{-10})} = 1.08 \times 10^3 \tag{G.15}$$

This calculation assumed that the intensity was spread over three lines that are 1 G wide, and assumed that the magnetic field modulation was chosen to observe all of the EPR line, rather than ca. 1/10 of the line.

Calculation of S/N for a Weak Pitch Sample

The sensitivity specification for a modern spectrometer is about 10^{10} spins per G linewidth. This value is based on a measurement of weak pitch using a slightly saturating microwave power and the modulation amplitude set larger than line width; conditions chosen to maximize the signal amplitude. The calculation of S/N for weak pitch assumes a line width broadened to 6 G by large modulation amplitude and 200 mW incident power.

The Varian catalog listed strong pitch as $3 \times 10^{15}\Delta$H spins/cm $\pm 15\%$. Yordanov and Ivanova (1994a) cited 3×10^{18} spins/cm for Bruker strong pitch, but this does not agree with other information. Weak pitch is diluted by a factor of 300 relative to strong pitch, so it has $1 \times 10^{13}\Delta$H spins/cm. The weak pitch spectrum is sharp in the center, with broad wings. Variations among samples are calibrated and the correction factor is used in the S/N test (see Appendix F). Line widths of weak pitch samples vary. To compensate for these variations, over-modulated spectra are used in signal-to-noise (S/N) tests, as described in Appendix F. To within the accuracy of the present discussion, the line width ΔH is estimated as 1.2 G. Hence, there are 1.2×10^{13} spins per cm of length. For a standard rectangular cavity, the active length of the resonator is about 2 cm, but with variation in signal over this length such that the effective signal increase is roughly 1.5 times that for a 1 cm sample. The effective filling factor (see table in Chap. 8) will thus be about 1.9% for a 3 mm i.d. weak pitch sample 2 cm long. This factor is used in the calculation of the signal voltage. Note that in denominator of (G.16), the experimental over-modulated peak-to-peak line width (6 G) is used and then converted to s^{-1}, because the spectrum is broadened by the modulation by roughly the same amount that the intensity is increased by the over-modulation (recall that the assumption is that the modulation is chosen such that all of the spin susceptibility is measured).

$$\begin{aligned}\chi'' &= (2.66 \times 10^{-32}) \frac{2\pi 9.4 \times 10^9}{(6)\sqrt{3}(2\pi 2.8 \times 10^6)} (2)(1.2 \times 10^{13} spins/0.14 \times 10^{-6} m^3)(0.8) \\ &= 1.2 \times 10^{-9}\end{aligned} \tag{G.16}$$

The signal voltage is calculated for 200 mW incident power, assuming that the signal does not saturate.

$$V_S = (1/2)(1.2 \times 10^{-9})(0.019)(3000)\sqrt{(50)(200 \times 10^{-3})}$$
$$= 1.08 \times 10^{-7} V \quad \text{(G.17)}$$

$$S/N = \frac{(1.08 \times 10^{-7})}{(1.58 \times 10^{-10})} = 684 \quad \text{(G.18)}$$

Thus, S/N = 684 is predicted for weak pitch in a perfect spectrometer with a TE_{102} cavity.

Note that the noise in the denominator is strongly dependent on the filter time constant and the equivalent noise effective bandwidth of the filter. Losses and the noise of active devices in the signal path might increase the noise relative to the EPR signal by 6 dB. Lacking an actual measurement, assume an effective overall noise figure, NF = 6 dB, which is a factor of 2 in noise voltage, resulting in a prediction of S/N = 342 for weak pitch. The Bruker specification for a standard rectangular TE_{102} cavity resonator on an Elexsys spectrometer is 400:1. For conditions comparable to the assumptions of this calculation, but with an optimized source and high-Q resonator on an EMXPlus spectrometer, Bruker now meets a 3000:1 S/N specification for weak pitch. Thus, this S/N calculation is fairly realistic, and is probably good to within a factor of two given the various approximations made, especially about filling factor and amount of the long line sample actually observed. The experimental weak pitch S/N measurement uses the derivative spectrum. In comparing the calculation with experiment, it is assumed that the noise floor is established prior to the phase-sensitive detector, and that the derivative spectrum is observed with the same S/N as the absorption.

Summary of Impact of Parameters on S/N

The following section discusses the contributions of noise other than thermal noise. First, review the parameters in the above equations. The signal voltage is given by (G.9). The properties of the sample dominate this – the spin concentration contributes linearly. The more spins the stronger the EPR signal. The narrower the line, the stronger the EPR signal for the same number of spins. The larger the sample, the larger the filling factor, but usually a larger sample lowers the Q, so there is a tradeoff. In addition, as pointed out in an earlier Chapter, the number of spins has to be kept small enough that the EPR signal is a small perturbation on the resonator Q. Increasing modulation amplitude increases S/N up to about an amplitude equal to the line width for narrow lines. Very large modulation amplitudes (more than a few gauss) may increase the noise level due to microphonics. Increasing gain will increase signal amplitude, but will not increase S/N. The other operator-controlled variables are the microwave frequency and the temperature, but these are usually dictated by the goals of the experiment and by the resources available.

This treatment ignores relaxation times. The electron spin relaxation time will determine the microwave power (P) that can be used. Operator judgment is required here also, trading off partial saturation for better S/N, as discussed in Sect. 4.11 and Appendix A.

How to Improve the Spectrometer: The Friis Equation

The estimates above used an overall noise figure for the detection system. One could measure each component to estimate its effect on the overall noise figure. To understand the importance of various losses and noise sources in a spectrometer, one would know all of the terms in the Friis equation. There are many ways to write the Friis equation. A way that is favored by engineers is in terms of the noise temperature of each stage. Noise temperature does not imply physical temperature.

$$T_{e1...n} = T_{e1} + \frac{T_{e2}}{g_1} + \cdots + \frac{T_{en}}{g_1 g_2 \cdots g_{n-1}} \tag{G.19}$$

where g_i is the power gain of the ith stage and T_{ei} is the noise temperature of the ith stage.

$T_e = T_0(F-1)$. The noise factor (F) is the ratio of the output noise power to the portion of the output noise power that is produced by the input thermal noise when at a standard temperature of 290 K. For a noiseless network $F = 1$ (the noise figure, NF, would be 0 dB) and $T_e = 0$. $T_0 = 290$ K. For practical calculations of a spectrometer, recall that a mixer, used as a phase-sensitive detector, has a loss of 1.44 dB (Rinard et al. 1999c). It can be seen from the Friis equation that the overall noise figure of a network is strongly dependent on the gain and NF of the first stage amplifier. If the early stages have low enough NF and high enough gain, the later stages are of little importance unless they introduce an enormous noise or interference signal (such as 60 Hz!). It is especially important to recognize that all losses prior to the first amplifier increase the effective noise figure by the amount of the losses.

Experimental Comparison

Although this seems rather forbidding, it is not impossible to make the necessary calculations and measurements. Starting with the number of spins in the sample, one calculates the signal as in the examples above, and compares it with the noise expected for the known (measured) gains, losses, and noise figures of each stage in the signal detection path of the spectrometer. A commercial CW EPR spectrometer is of course the hardest case, because one does not know the gains and losses and noise figures of all of the components in the signal detection path. Consequently, results are given for a locally-constructed pulsed EPR spectrometer. For a particular case an S-band spin echo was calculated to be 3.0 V at the digital oscilloscope, and the measured echo was 2.9 V (Rinard et al. 1999c).

These results provide confidence that the approach used is a valid way to identify the major targets for improvement of EPR spectrometer performance. Immediately evident from the Friis equation is that loss prior to the first stage amplifier, and the noise figure of the first stage amplifier can dominate the overall noise figure of the system. In older CW EPR spectrometers, the detector and first stage amplifier had poorer noise figures than is now possible due to technology improvements. From the discussions of η, Q, B_1, and modulation amplitude it is evident that resonators with higher Q and more uniform B_1, and modulation coil systems with more uniform modulation over the sample, all could yield improved signal amplitudes. The Bruker "high-Q" resonators exploit some of these features to provide the user with improved S/N. When everything else is improved, microwave source noise becomes an important contributor. Hence, for the highest-Q resonators, the microwave source has to be specially selected to have the lowest noise among those currently available.

Sometimes, it is necessary to trade off S/N for other experimental goals. For example, there is almost no loss in short lengths of X-band waveguide, but there is not enough room in the normal cryostat that fits in a normal magnet gap to use waveguide to the cavity. Hence, in cryostats coaxial cable is used at X-band (at Q-band, waveguide can be used). A meter of coaxial cable results in a loss of about 16% of the signal voltage (ca. 1.5 dB) at X-band.

The Friis equation points to another tradeoff. A low-noise preamplifier could greatly improve the EPR S/N by establishing the noise floor before significant losses, such as in the detector crystal. However, a low-noise preamplifier will not survive more than a few milliwatts incident power. Consequently, if a low-noise preamplifier is used in a general-purpose spectrometer, the operator has to switch it out of the circuit whenever the power limits of the amplifier would be exceeded, or it has to be protected by a limiter, which itself increases the noise figure. There is no free lunch. A low-noise preamplifier is used in the Bruker E580 pulsed spectrometer, where it is protected by an active limiter (microwave switch) and can be switched out of the circuit by the operator. If, however, a low-noise preamplifier could be installed at the output of the resonator, as might be possible with a bimodal resonator, a factor of at least two in S/N performance might be possible.

References

Aasa, R., Vanngard, T.: EPR signal intensity and powder shapes. Reexamination. J. Magn. Reson. **19**, 308–315 (1975)

Abragam, A.: The Principles of Nuclear Magnetism. Oxford University Press, Oxford (1961)

Adem, E., Munoz, P.E., Gleason, V.R., Murrieta, S.H., Aguilar, S.G., Uribe, R.E.: Electron paramagnetic resonance studies of γ-irradiated corn. Appl. Radiat. Isot. **44**, 419–422 (1993)

Alaouie, A.M., Smirnov, A.I.: Ultra-stable temperature control in EPR experiments: thermodynamics of gel-to-liquid phase transition in spin-labeled phospholipid bilayers and bilayer perturbations by spin labels. J. Magn. Reson. **182**, 229–238 (2006)

Alger, R.S.: Electron Paramagnetic Resonance: Techniques and Applications. Wiley, New York (1968)

Anderson, W.A., Piette, L.H.: Forbidden $m_s = \pm 1$, $m_I = \pm 1$ transitions in a vanadyl chelate. J. Chem. Phys. **30**, 591–592 (1959)

Anderson, J.R., Mett, R.R., Hyde, J.S.: Cavities with axially uniform fields for use in electron paramagnetic resonance. II. Free space generalization. Rev. Sci. Instrum. **73**(8), 3027–3037 (2002)

Ardenkjaer-Larsen, J.H., Laursen, I., Leunbach, I., Ehnholm, G., Wistrand, L.-G., Petersson, J.S., Golman, K.: EPR and DNP properties of certain novel single electron contrast agents intended for oximetric imaging. J. Magn. Reson. **133**, 1–12 (1998)

Ashton, G.R., Hsu, D.K., Leisure, R.G.: Comparison of multiple-scan direct and lock-in detection in magnetic resonance: application to nuclear acoustic resonance. Rev. Sci. Instrum. **51**, 454–458 (1980)

Atherton, N.M.: Principles of Electron Spin Resonance. Ellis Horwood Prentice Hall, London (1993)

Auteri, F.P., Belford, R.L., Boyer, S., Motsegood, K.K., Smirnov, A.I., Smirnova, T., Vahidi, N., Clarkson, R.B.: Carbon-based standards for electron paramagnetic resonance spectroscopy. Appl. Magn. Reson. **6**, 287–308 (1994)

Baberschke, K., Bures, K.D., Barnes, S.E.: ESR in situ with a Josephson tunnel junction. Phys. Rev. Lett. **53**, 98–101 (1984)

Balan, E., Allard, T., Boizot, B., Morin, G., Muller, J.-P.: Quantitative measurement of Fe^{3+} in kaolinite. Clays Clay Miner. **48**, 439–445 (2000)

Bales, B.L., Kevan, L.: Paramagnetic relaxation of silver species in γ-irradiated frozen aqueous solutions. J. Chem. Phys. **52**, 4644–4653 (1970)

Belford, R.L., Clarkson, R.B., Cornelius, J.B., Rothenberger, K.S., Nilges, M.J., Timken, M.D.: EPR over three decades of frequency: radiofrequency to infrared. In: Weil, J.A., Bowman, M.K., Preston, K.F. (eds.) Electron Magnetic Resonance of the Solid State, pp. 21–43. Canadian Society for Chemistry, Ottawa (1987)

Bender, C.J.: ENDOR coils and related radiofrequency circuits. In: Bender, C.J., Berliner, L.J. (eds.) EPR: Instrumental Methods. Biological Magnetic Resonance, vol. 21, pp. 155–211. Kluwer Academic/Plenum, New York (2004)

Bertini, I., Martini, G., Luchinat, C.: Relaxation data tabulation. In: Poole Jr., C.P., Farach, H. (eds.) Handbook of Electron Spin Resonance: Data Sources, Computer Technology, Relaxation, and ENDOR, pp. 79–310. American Institute of Physics, New York (1994a)

Bertini, I., Martini, G., Luchinat, C.: Relaxation, background, and theory. In: Poole Jr., C.P. (ed.) Handbook of Electron Spin Resonance, pp. 51–77. American Institute of Physics, New York (1994b)

Beth, A., Hustedt, E.J.: Saturation transfer EPR: rotational dynamics of membrane proteins. In: Eaton, S.S., Eaton, G.R., Berliner, L.J. (eds.) Biomedical EPR, Part B: Methodology, Instrumentation, and Dynamics. Biological Magnetic Resonance, vol. 24, pp. 369–407. Kluwer Academic/Plenum, New York (2005)

Beth, A., Balasubramanian, K., Robinson, B.H., Dalton, L.R., Venkataramu, S.K., Park, J.H.: Sensitivity of V2′ saturation transfer electron paramagnetic resonance signals to anisotropic rotational diffusion with [15N]nitroxide spin-labels. Effects of noncoincident magnetic and diffusion tensor principal axes. J. Phys. Chem. **87**, 359–367 (1983)

Blakley, R.L., Henry, D.D., Morgan, W.T., Clapp, W.L., Smith, C.J., Barr, D.: Quantitative electron paramagnetic resonance: the importance of matching the Q-factor of standards and samples. Appl. Spectrosc. **55**, 1375–1381 (2001)

Bloembergen, N., Pound, R.V.: Radiation damping in magnetic resonance experiments. Phys. Rev. **95**, 8–12 (1954)

Bloembergen, N., Purcell, E.M., Pound, R.V.: Relaxation effects in nuclear resonance absorption. Phys. Rev. **73**, 679–712 (1948)

Bolton, J.R., Borg, D.C., Swartz, H.M.: Experimental aspects of biological electron spin resonance studies. In: Swartz, H.M., Bolton, J.R., Borg, D.C. (eds.) Biological Applications of Electron Spin Resonance. Wiley, New York (1972). Ch. 2

Box, H.C.: Radiation Effects: ESR and ENDOR Analysis. Academic, New York (1977)

Brändel, R., Krüger, G.J., Müller-Warmuth, W.: Impulsspektroskopische Untersuchungen der Elektronenspinrelaxation in freien Radikalen. Z. Naturforsch. **25a**, 1–11 (1970)

Bryson, W.G., Hubbard, D.P., Peake, B.M., Simpson, J.: Applications of electron spin resonance in the analytical chemistry of transition metal ions. Part I. Factors affecting the signal amplitude. Anal. Chim. Acta **77**, 107–115 (1975)

Budil, D.E., Lee, S., Saxena, S., Freed, J.H.: Nonlinear least-squares analysis of slow-motion EPR spectra in one and two dimensions using a modified Levenberg-Marquardt algorithm. J. Magn. Reson. A **120**, 155–189 (1996)

Burgess, J.H., Brown, R.M.: Modulation effects in nuclear magnetic resonance. Rev. Sci. Instrum. **23**, 334–336 (1952)

Burns, D.T., Flockhart, B.D.: Application of quantitative EPR. Philos. Trans. R. Soc. Lond. A **333**, 37–48 (1990)

Cage, B., Cevc, P., Blinc, R., Brunel, L.C., Dalal, N.S.: 1-370 GHz EPR linewidths for K_3CrO_8: a comprehensive test for the Anderson-Weiss model. J. Magn. Reson. **135**, 178–184 (1998)

Calvo, R., Abresch, E.C., Bittl, R., Feher, G., Hofbauer, W., Isaacson, R.A., Lubitz, W., Okamura, M.Y., Paddock, M.L.: EPR study of the molecular and electronic structure of the semiquinone biradical. Q_A^-. Q_B^-. in photosynthetic reaction centers from Rhodobacter sphaeroides. J. Am. Chem. Soc. **122**, 7327–7341 (2000)

Carl, P., Höfer, P.: Quantitative EPR techniques. Bruker Spin Rep. **159**(160), 17–21 (2008)

Carrington, A., Luckhurst, G.R.: Electron spin resonance line widths of transition metal ions in solution. Relaxation through zero-field splitting. Mol. Phys. **8**, 125–132 (1964)

Casteleijn, G., TenBosch, J.J., Smidt, J.: Error analysis of the determination of spin concentration with the electron spin resonance method. J. Appl. Phys. **39**, 4375–4380 (1968)

Chang, R.: Simple setup for quantitative electron paramagnetic resonance. Anal. Chem. **46**, 1360 (1974)

Chang, T.-T., Foster, D., Kahn, A.H.: An intensity standard for electron paramagnetic resonance using chromium-doped corundum (Al_2O_3: Cr^{3+}). J. Res. Natl. Bur. Stand. **83**, 133–164 (1978)

Chesnut, D.B.: On the use of AW2 method for integrated line intensities from first-derivative presentations. J. Magn. Reson. **25**, 373–374 (1977)

Chien, J.C.W., Wang, D.S.T.: Autoxidation of Polyolefins. Absolute Rate Constants and Effect of Morphology. Macromolecules **8**, 920–928 (1975)

Conradi, M.S.: Magnetic resonance signal-to-noise calculations: direct vs. lock-in recording. Rev. Sci. Instrum. **48**, 444–448 (1977)

Copeland, E.S.: Simple method for estimating H_1 [microwave magnetic field strength] in ESR experiments. Microwave power saturation of γ-irradiation induced glycylglycine radicals. Rev. Sci. Instrum. **44**, 437–442 (1973)

Cordischi, D., Occhiuzzi, M., Dragone, R.: Quantitative EPR spectroscopy: comparison between primary standards and application to MgO-MnO and α-Al_2O_3-CrO_3 solid solutions. Appl. Magn. Reson. **16**, 427–445 (1999)

Crippa, P.R., Urbinati, E.U., Vecli, A.: On the determination of the unpaired spin number by electron spin resonance. J. Phys. E. **4**, 1071–1073 (1971)

Czoch, R.: Quantitative EPR – sensitivity to experimental conditions and optimal setting of recording parameters. Appl. Magn. Reson. **10**, 293–317 (1996)

Dalal, D.P., Eaton, S.S., Eaton, G.R.: The effects of lossy solvents on quantitative EPR studies. J. Magn. Reson. **44**(3), 415–428 (1981a)

Dalal, N.S., Suryan, M.M., Seehra, M.S.: Potassium perchromate standard for determination of para-magnetic spin concentration, g values, and magnetic moments of fossil fuels. Anal. Chem. **53**, 938–940 (1981b)

Dalton, L.R., Robinson, B.H., Dalton, L.A., Coffey, P.: Saturation transfer spectroscopy. Adv. Magn. Reson. **8**, 149–259 (1976)

Dashnau, J.L., Zelent, B., Vanderkooi, J.M.: Tryptophan interactions with glycerol/water and trehalose/sucrose cryosolvents: infrared and fluorescence spectroscopy and ab initio calculations. Biophys. Chem. **114**, 71–83 (2005)

Devine, S.D., Robinson, W.H.: Ultrasonically modulated paramagnetic resonance. Adv. Magn. Reson. **10**, 53–117 (1982)

Drago, R.S.: Physical Methods in Chemistry. Saunders, USA (1977)

Drew, S.C., Young, C.G., Hanson, G.R.: A density functional study of the electronic structure and spin Hamiltonian parameters of mononuclear thiomolybdenyl complexes. Inorg. Chem. **46**, 2388–2397 (2007)

Dyrek, K., Madej, A., Mazur, E., Rokosz, A.: Standards for EPR measurements of spin concentration. Colloids Surf. **45**, 135–144 (1990)

Dyrek, K., Rokosz, A., Madej, A.: Spin dosimetry in catalysis research. Appl. Magn. Reson. **6**, 309–332 (1994)

Dyrek, K., Rokosz, A., Madej, A., Bidzinska, E.: Quantitative EPR studies of transition metal ions in oxide, aluminosilicate and polymer matrices. Appl. Magn. Reson. **10**, 319–338 (1996)

Dyrek, K., Bidzinska, E., Adamski, A.: Quantitative EPR – a versatile tool in fundamental and applied studies. Mol. Phys. Rep. **37**, 9–23 (2003)

Eachus, R.S., Olm, M.T.: Electron nuclear double resonance spectroscopy. Science **230**, 268–274 (1985)

Eastman, M.P., Kooser, R.G., Das, M.R., Freed, J.H.: Heisenberg spin exchange in E.S.R. spectra. I. Linewidth and saturation effects. J. Chem. Phys. **51**, 2690–2709 (1969)

Eaton, S.S., Eaton, G.R.: Electron paramagnetic resonance cell for lossy samples. Anal. Chem. **49**, 1277–1278 (1977)

Eaton, S.S., Eaton, G.R.: Signal area measurements in EPR. Bull. Magn. Reson. **1**(3), 130–138 (1980)

Eaton, S.S., Eaton, G.R.: Measurement of spin-spin distances from the intensity of the EPR half-field transition. J. Am. Chem. Soc. **104**(18), 5002–5003 (1982)

Eaton, S.S., Eaton, G.R.: EPR imaging. Spectroscopy **1**(1), 32–35 (1986)

Eaton, G.R., Eaton, S.S.: Electron paramagnetic resonance. In: Ewing 2nd, G.W. (ed.) Analytical Instrumentation Handbook, pp. 467–530. Marcel Dekker, New York (1990)

Eaton, S.S., Eaton, G.R.: Quality assurance in EPR. Bull. Magn. Reson. **13**(3–4), 83–89 (1992)

Eaton, S.S., Eaton, G.R.: Irradiated fused-quartz standard sample for time-domain EPR. J. Magn. Reson. **102**(3), 354–356 (1993)

Eaton, S.S., Eaton, G.R.: Electron paramagnetic resonance. In: Ewing, G.W. (ed.) Analytical Instrumentation Handbook, 2nd edn, pp. 767–862. Marcel Dekker, New York (1997)

Eaton, S.S., Eaton, G.R.: EPR imaging. Electron Paramagn. Reson. **17**, 109–129 (2000a)

Eaton, S.S., Eaton, G.R.: Relaxation times of organic radicals and transition metal ions. In: Berliner, L.J., Eaton, S.S., Eaton, G.R. (eds.) Distance Measurements in Biological Systems by EPR. Biological Magnetic Resonance, vol. 19, pp. 29–154. Kluwer Academic/Plenum, New York (2000b)

Eaton, S.S., Eaton, G.R.: Electron paramagnetic resonance. In: Cazes, J. (ed.) Analytical Instrumentation Handbook, 3rd edn, pp. 349–398. Marcel Dekker, New York (2005)

Eaton, S.S., Law, M.L., Peterson, J., Eaton, G.R., Greenslade, D.J.: Metal-nitroxide interactions. 7. Quantitative aspects of EPR spectra resulting from dipolar interactions. J. Magn. Reson. **33**(1), 135–141 (1979)

Eaton, S.S., More, K.M., Sawant, B.M., Eaton, G.R.: Use of the ESR half-field transition to determine the interspin distance and the orientation of the interspin vector in systems with two unpaired electrons. J. Am. Chem. Soc. **105**(22), 6560–6567 (1983)

Eaton, G.R., Eaton, S.S., Ohno, K. (eds.): EPR Imaging and In Vivo EPR. CRC Press, Boca Raton, FL (1991)

Eaton, G.R., Eaton, S.S., Rinard, G.A.: Frequency dependence of EPR sensitivity. In: Spatially resolved magnetic resonance: methods, materials, medicine, biology, rheology, geology, ecology, hardware. Blümler, P., Blümich, B., Bottto, R.E., Fukushima, E. (eds.) Wiley-VCH, Weinheim. Based on lectures presented at the Fourth International Conference on Magnetic Resonance Microscopy, Albuquerque, Oct 1997, pp. 65–74 (1998)

Eaton, G. R., Eaton, S. S., Salikhov, K. M. (eds): Foundations of Modern EPR. World Scientific, Singapore (1998)

Eaton, S.S., Harbridge, J., Rinard, G.A., Eaton, G.R., Weber, R.T.: Frequency dependence of electron spin relaxation for three S = 1/2 species doped into diamagnetic solid hosts. Appl. Magn. Reson. **20**(1–2), 151–157 (2001)

Elas, M., Ahn, K.H., Parasca, A., Barth, E.D., Lee, D., Haney, C., Halpern, H.J.: Electron paramagnetic resonance oxygen images correlate spatially and quantitatively with OxyLite oxygen measurements. Clin. Cancer Res. **12**, 4209–4217 (2006)

Emge, T.J., Wang, H.H., Beno, M.A., Leung, P.C.W., Firestone, M.A., Jenkins, H.C., Cook, J.D., Carlson, K.D., Williams, J.M., Venturini, E.L., Azevedo, J., Schirber, J.E.: A test of superconductivity vs. molecular disorder in $(BEDT\text{-}TTF)_2X$ synthetic metals: synthesis, structure (298, 120 K), and microwave/ESR conductivity of $(BEDT\text{-}TTF)_2I_2Br$. Inorg. Chem. **24**, 1736–1738 (1985)

Farrar, T.C., Becker, E.D.: Pulse and Fourier Transform NMR. Academic, New York (1971)

Fedin, M.V., Purtov, P.A., Bagryanskaya, E.G.: Spin relaxation in low and zero magnetic field. J. Chem. Phys. **118**, 192–201 (2003a)

Fedin, M.V., Purtov, P.A., Bagyanskaya, E.G.: Spin relaxation of radicals in low and zero magnetic field. J. Chem. Phys. **118**, 192–201 (2003b)

Feher, G.: Sensitivity considerations in microwave paramagnetic resonance absorption techniques. Bell Syst. Tech. J. **36**, 449–484 (1957)

Freed, J.H.: Theory of slow tumbling ESR spectra of nitroxides. In: Berliner, L.J. (ed.) Spin Labeling: Theory and Applications, pp. 53–132. Academic, New York (1976)

Froncisz, W., Oles, T., Hyde, J.S.: Q-band loop-gap resonator. Rev. Sci. Instrum. **57**(6), 1095–1099 (1986)

Fuchs, M., Groth, N., Herrling, T.: Application of in vivo EPR spectroscopy and imaging to skin. In: Berliner, L.J. (ed.) In Vivo EPR(ESR). Biological Magnetic Resonance, vol. 18, pp. 483–515. Kluwer Academic/Plenum, New York ((2003)

Fujita, Y., Tsuchiya, K., Abe, S., Takiguchi, Y., Kubo, S., Sakurai, H.: Estimation of the age of human bloodstains by electron paramagnetic resonance spectroscopy: long-term controlled experiment on the effects of environmental factors. Forensic Sci. Int. **152**, 39–43 (2005)

Gancheva, V., Yordanov, N.D., Callens, F., Vanhaelewyn, G., Raffi, J., Bortolin, E., Onori, S., Malinen, E., Sagstuen, E., Fabisiak, S., Peimel-Stuglik, Z.: An international intercomparison on 'self-calibrated' alanine dosimeters. Radiat. Phys. Chem. **77**, 357–364 (2008)

Gerson, F., Gescheidt, G.: Solution-ENDOR of some biologically interesting radical ions. In: Eaton, S.S., Eaton, G.R., Berliner, L.J. (eds.) Biomedical EPR – Part B: Methodology, Instrumentation, and Dynamics. Biological Magnetic Resonance, vol. 24, pp. 145–164. Kluwer Academic/Plenum, New York (2005)

Ginzton, E.L.: Microwave Measurements. McGraw-Hill, New York (1957)

Glarum, S.H., Marshall, J.H.: Ethereal electrons. J. Chem. Phys. **52**, 5555–5565 (1970)

Goldberg, I.B.: Improving the analytical accuracy of electron paramagnetic resonance spectroscopy. J. Magn. Reson. **32**, 233–242 (1978)

Goldberg, I.B., Bard, A.J.: Electron-spin-resonance spectroscopy. In: Kolthoff, I.M., Elving, P.J. (eds.) Treatise on Analytical Chemistry, vol. 10, Part I, 2nd edn, pp. 226–289. Wiley, New York (1983)

Goldberg, I.B., Crowe, H.R.: Effect of cavity loading on analytical electron spin resonance spectrometry. Anal. Chem. **49**, 1353–1357 (1977)

Goldberg, I.B., Crowe, H.R., Robertson, W.M.: Determination of the composition of mixtures of sodium chromite and chromic oxide by electron spin resonance spectrometry. Anal. Chem. **49**, 962–966 (1977)

Golumbek, A.P., Hendrich, M.P.: Quantitative analysis of dinuclear manganese(II) EPR spectra. J. Magn. Reson. **165**, 33–48 (2003)

Grinberg, O., Berliner, L.J. (eds.): Very High Frequency (VHF) ESR/EPR. Biological Magnetic Resonance, vol. 22. Kluwer Academic/Plenum, New York (2004)

Grucker, D., Guiberteau, T., Planinšic, G.: Proton-electron double resonance: spectroscopy and imaging in very low magnetic fields. Res. Chem. Intermed. **22**, 567–579 (1996)

Guiberteau, T., Grucker, D.: Dynamic nuclear polarization of water protons by saturation of σ and π EPR transitions of nitroxides. J. Magn. Reson. A. **105**, 98–103 (1993)

Guiberteau, T., Grucker, D.: EPR spectroscopy by dynamic nuclear polarization in low magnetic field. J. Magn. Reson. B **110**, 47–54 (1996)

Gunter, T.E., Puskin, J.S., Russell, P.R.: Quantitative magnetic resonance studies of manganese uptake by mitochondria. Biophys. J. **15**, 319–333 (1975)

Hagen, W.R., Wassink, H., Eady, R.R., Smith, B.E., Haaker, H.: Quantitative EPR of an $S = 7/2$ system in thionine-oxidized MeFe proteins of nitrogenase. A redefinition of the P-cluster concept. Eur. J. Biochem. **169**, 457–465 (1987)

Halpern, H.J.: Applications of in vivo EPR spectroscopy and imaging in cancer research. In: Berliner, L.J. (ed.) In Vivo EPR (ESR). Biological Magnetic Resonance, vol. 18, pp. 469–482. Kluwer Academic/Plenum, New York (2003)

Halpern, H.J., Peric, M., Yu, C., Bales, B.L.: Rapid quantitation of parameters from inhomogeneously broadened EPR spectra. J. Magn. Reson. A **103**(1), 13–22 (1993)

Halpern, H.J., Yu, C., Peric, M., Barth, E., Grdina, D.J., Teicher, B.A.: Oxymetry deep in tissues with low-frequency electron paramagnetic resonance. Proc. Natl. Acad. Sci. U.S.A. **91**(26), 13047–13051 (1994)

Hanson, G., Szabo, A., Chasteen, N.D.: Determination of Molybdenum in Seawater by Electron Paramagnetic Resonance Spectroscopy. Anal. Chem. **49**, 461–463 (1977)

Harbridge, J.R., Eaton, G.R., Eaton, S.S.: Impact of spectral diffusion on apparent relaxation times for the stable radical in irradiated glycylglycine. In: Modern applications of EPR/ESR: from biophysics to materials science. In: Proceedings of the First Asia-Pacific EPR/ESR Symposium, Kowloon, Hong Kong, 20–24 Jan 1997, pp. 220–225 (1998)

Harbridge, J.R., Eaton, S.S., Eaton, G.R.: Electron spin-lattice relaxation processes of radicals in irradiated crystalline organic compounds. J. Phys. Chem. A **107**(5), 598–610 (2003)

Hemminga, M.A., deJager, P.A., Marsh, D., Fajer, P.: Standard conditions for the measurement of saturation-transfer ESR spectra. J. Magn. Reson. **59**, 160 (1984)

Hendrich, M.P., DeBrunner, P.G.: Integer-spin electron paramagnetic resonance of iron proteins. Biophys. J. **56**, 489–506 (1989)

Hernandez, A., Martin, E., Margineda, J., Zamarro, J.M.: Resonant cavities for measuring the surface resistivities of metals at X-band frequencies. J. Phys. E. **19**, 222–225 (1986)

Herrington, J.R., Boatner, L.A., Aton, T.J., Estle, T.L.: Electron paramagnetic resonance investigation of the dynamic Jahn-Teller effect for Sc^{2+} in BaF_2, SrF_2, and CaF_2. Phys. Rev. B **10**, 833–843 (1974)

Hill, H.D.W., Richards, R.E.: Limits of measurement in magnetic resonance. J. Sci. Instrum. **2**, 977–983 (1968)

Horasan, N., Sünnetçioğlu, M.M., Sungur, R., Bingöl, G.: A weak field EPR study of TANOL in water and water-oil using dynamic nuclear polarization. Z. Naturforsch. **52a**, 485–489 (1997)

Hyde, J.S.: A new principle for aqueous sample cells for EPR. Rev. Sci. Instrum. **43**, 629–631 (1972)

Hyde, J.S.: Saturation-transfer spectroscopy. Meth. Enzymol. **49**, 480–511 (1978)

Hyde, J.S., Froncisz, W.: The role of microwave frequency in EPR spectroscopy of copper complexes. Annu. Rev. Biophys. Bioeng. **11**, 391–417 (1982)

Hyde, J.S., Mett, R.R.: Aqueous sample considerations in uniform field resonators for electron paramagnetic resonance spectroscopy. Curr. Top. Biophys. **26**, 7–14 (2002)

Hyde, J.S., Thomas, D.D.: Saturation-transfer spectroscopy. Annu. Rev. Phys. Chem. **31**, 293–317 (1980)

Hyde, J.S., Jesmanowicz, A., Ratke, J.J., Antholine, W.E.: Pseudomodulation: a computer-based strategy for resolution enhancement. J. Magn. Reson. **96**(1), 1–13 (1992)

Hyde, J.S., Mett, R.R., Anderson, J.R.: Cavities with axially uniform fields for use in electron paramag-netic resonance. III. Re-entrant geometries. Rev. Sci. Instrum. **73**(11), 4003–4009 (2002)

Ilangovan, G., Zweier, J.L., Kuppusamy, P.: Mechanism of oxygen-induced EPR line broadening in lithium phthalocyanine microcrystals. J. Magn. Reson. **170**, 42–48 (2004)

Jardetzky, O., Roberts, G.C.K.: NMR in Molecular Biology. Academic, New York (1981)

Jeschke, G.: Instrumentation and Experimental Setup. Biol. Magn. Reson. **27**, 17–47 (2007)

Joshi, J.P., Ballard, J.R., Rinard, G.A., Quine, R.W., Eaton, G.R.: Rapid-scan EPR with triangular scans and Fourier deconvolution to recover the slow-scan spectrum. J. Magn. Reson. **175**, 44–51 (2005a)

Joshi, J.P., Eaton, G.R., Eaton, S.S.: Impact of resonator on direct-detected rapid-scan EPR at 9.8 GHz. Appl. Magn. Reson. **29**, 239–249 (2005b)

Juarez-Garcia, C., Hendrich, M.P., Holman, T.R., Que, L.J., Münck, E.: Combined mossbauer and EPR studies of the S = 3 state of an exchange-coupled FeIIICuII complex: test for quantitative EPR analysis of integer spin systems. J. Am. Chem. Soc. **113**, 518–525 (1991)

Kälin, M., Gromov, I., Schweiger, A.: The continuous wave electron paramagnetic resonance experiment revisited. J. Magn. Reson. **160**, 166–182 (2003)

Kevan, L., Kispert, L.D.: Electron Spin Double Resonance Spectroscopy. Wiley, New York (1976)

Kirste, B., Harrer, W., Kurreck, H.: ENDOR studies of novel Di- and triphenylmethyl radicals generated from galvinols. J. Am. Chem. Soc. **107**, 20–28 (1985)

Kispert, L.D.: Electron-electron double resonance. In: Eaton, S.S., Eaton, G.R., Berliner, L.J. (eds.) Biomedical EPR – Part B: Methodology, Instrumentation, and Dynamics. Biological Magnetic Resonance, vol. 24, pp. 165–197. Kluwer Academic/Plenum, New York (2005)

Kolaczkowski, S.V., Cardin, J.T., Budil, D.E.: Some remarks on reported inconsistencies in the high-field EPR spectrum of DPPH. Appl. Magn. Reson. **16**, 293–298 (1999)

Kooser, R.G., Volland, W.V., Freed, J.H.: E.S.R. relaxation studies on orbitally degenerate free radicals. I. Benzene anion and tropenyl. J. Chem. Phys. **50**, 5243–5257 (1969)

Kooser, R.G., Kirchman, E., Matkov, T.: Measurements of spin concentration in electron paramagnetic resonance spectroscopy. Concepts Magn. Reson. **4**, 145–152 (1992)

Kuppusamy, P., Chzhan, M., Zweier, J.L.: Principles of imaging. In: Berliner, L.J. (ed.) In Vivo EPR (ESR). Biological Magnetic Resonance, vol. 18, pp. 99–152. Kluwer Academic/Plenum, New York (2003)

Kurreck, H., Bock, M., Bretz, N., Elsner, M., Lubitz, W., Muller, F., Geissler, J., Kroneck, P.M.H.: Fluid solution and solid-state electron nuclear double resonance of flavin model compounds and flavoenzymes. J. Am. Chem. Soc. **106**, 737–746 (1984)

Lloyd, J.P., Pake, G.E.: Spin relaxation in free radical solutions exhibiting hyperfine structure. Phys. Rev. **94**, 579–591 (1954)

Ludowise, P., Eaton, S.S., Eaton, G.R.: A convenient monitor of EPR automatic frequency control function. J. Magn. Reson. **93**(2), 410–412 (1991)

Lurie, D.J.: Proton-electron double-resonance imaging (PEDRI). In: Berliner, L.J. (ed.) In Vivo EPR(ESR). Biological Magnetic Resonance, vol. 18, pp. 547–578. Kluwer Academic/Plenum, New York (2003)

Lurie, D.J., Nicholson, I., Mallard, J.R.: Low-field EPR measurements by field-cycled dynamic nuclear polarization. J. Magn. Reson. **95**, 405–409 (1991)

Mailer, C., Sarna, T., Swartz, H.M., Hyde, J.S.: Quantitative studies of free radicals in biology: corrections to ESR saturation data. J. Magn. Reson. **25**, 205–210 (1977)

Mailer, C., Robinson, B.H., Williams, B.B., Halpern, H.J.: Spectral fitting: the extraction of crucial information from a spectrum and a spectral image. Magn. Reson. Med. **49**, 1175–1180 (2003)

Maresch, G.G., Mehring, M., von Schutz, J.U., Wolf, H.C.: Time-resolved ESR investigation of electron-spin diffusion in the radical cation salt $(\text{fluoranthenyl})_2{}^{+}AsF_6{}^{-}$. Chem. Phys. Lett. **85**, 333–340 (1984)

Marsh, D., Horváth, L.I., Páli, T., Livshits, V.: Saturation transfer spectroscopy of biological membranes. Biol. Magn. Reson. **24**, 309–367 (2005)

Mazúr, M.: A dozen useful tips on how to minimize the influence of sources of error in quantitative electron paramagnetic resonance (EPR) spectroscopy – a review. Anal. Chim. Acta **561**, 1–15 (2006)

Mazúr, M., Valko, M.: Radial effect of the EPR signal intensity in a Bruker single TE_{102} rectangular cavity. Bruker Rep **147**, 43–45 (1999)

Mazúr, M., Valko, M., Klement, R., Morris, H.: Quantitative electron paramagnetic resonance (EPR) spectrometry with a TE_{104} double rectangular cavity. Part 1. A simple alignment procedure for the precision positioning of the sample. Anal. Chim. Acta **333**, 249–252 (1996a)

Mazúr, M., Valko, M., Morris, H., Klement, R.: Quantitative electron paramagnetic resonance (EPR) spectrometry with a TE_{104} double rectangular cavity. Part 2. Analysis of sample and TE_{104} cavity error sources associated with the movement of line-like samples into the TE_{104} cavity. Anal. Chim. Acta **333**, 253–265 (1996b)

Mazúr, M., Morris, H., Valko, M.: Analysis of the movement of line-like samples of variable length along the x-axis of a double TE_{104} and a single TE_{102} rectangular resonator. J. Magn. Reson. **129**, 188–200 (1997a)

Mazúr, M., Valko, M., Morris, H.: A simple alignment procedure for the precision positioning of the sample at arbitrary points of the intracavity space of a TE_{104} double rectangular cavity. Rev. Sci. Instrum. **68**, 2514–2517 (1997b)

Mazúr, M., Valko, M., Pelikan, P.: Quantitative EPR spectroscopy in solid state chemistry. Chem. Pap. **51**, 134–136 (1997c)

Mazúr, M., Valko, M., Micov, M., Morris, H.: Analytical solution for the electron paramagnetic resonance signal intensity of a line-like sample of variable length whose centre is situated at an arbitrary position along the common sample-cavity x-axis. Anal. Chim. Acta **373**, 107–109 (1998)

Mazúr, M., Valko, M., Morris, H.: Analysis of the radial and longitudinal effect in a double TE_{104} and a single TE_{102} rectangular cavity. J. Magn. Reson. **142**, 37–56 (2000)

Mazúr, M., Valko, M., Morris, H.: Influence of the movement of "over full-length cavity" cylindrical samples along the x-axis of a double TE_{104} and a single TE_{102} rectangular cavity on the electron paramagnetic resonance. An unusual effect analysis. Anal. Chim. Acta **443**, 127–141 (2001)

Mazúr, M., Valko, M., Morris, H.: Influence of the variable wall thickness of the sample tube and a quartz Dewar on an EPR signal intensity in a single TE_{102} and double TE_{104} rectangular cavities. Anal. Chim. Acta **482**, 229–248 (2003)

Mazúr, M., Moncol, J., Valko, M., Morris, H.: Analysis of the radial and longitudinal effect of a planar sample in a single TE_{102} rectangular electron paramagnetic resonance (EPR) cavity. Anal. Chim. Acta **526**, 163–176 (2004)

Mazúr, M., Moncol, J., Valko, M., Morris, H.: Analysis of the longitudinal "sloping plateau" effect in a single TE_{102} rectangular cavity. Anal. Chim. Acta **538**, 165–174 (2005)

Mett, R.R., Hyde, J.S.: Aqueous flat cells perpendicular to the electric field for use in electron paramagnetic resonance spectroscopy. J. Magn. Reson. **165**, 137–152 (2003)

Mett, R.R., Froncisz, W., Hyde, J.S.: Axially uniform resonant cavity modes for potential use in electron paramagnetic resonance spectroscopy. Rev. Sci. Instrum. **72**(11), 4188–4200 (2001)

Mims, W.B.: The Linear Electric Field Effect in Paramagnetic Resonance. Clarendon, Oxford (1976)

Möbius, K., Savitsky, A.N.: High-Field EPR Spectroscopy on Proteins and Their Model Systems. Royal Society of Chemistry, Cambridge (2009)

Molin, Y.N., Chibrikin, V.M., Shabalkin, V.A., Shuvalov, V.F.: Accuracy of the E.P.R. (electron paramagnetic resonance) measurement of paramagnetic-particles concentration. Zavod. Lab. **32**, 933 (1966)

More, K.M., Eaton, G.R., Eaton, S.S.: Magnetic susceptibility and EPR changes caused by grinding of samples. J. Magn. Reson. **37**(2), 217–222 (1980)

More, K.M., Eaton, G.R., Eaton, S.S.: Determination of T_1 and T_2 by simulation of EPR power saturation curves and saturated spectra. Application to spin-labeled iron porphyrins. J. Magn. Reson. **60**(1), 54–65 (1984)

Morrisett, J.D.: Spin labeled enzymes. In: Berliner, L.J. (ed.) Spin Labeling: Theory and Applications, vol. I, p. 293. Academic, New York (1976)

Nagy, V.: Quantitative EPR: some of the most difficult problems. Appl. Magn. Reson. **6**, 259–285 (1994)

Nagy, V.Y.: Choosing reference samples for electronic paramagnetic resonance (EPR) spectrometric concentration measurements. Part 1. General introduction and systems of S = 1/2. Anal. Chim. Acta **339**, 1–11 (1997)

Nagy, V.Y., Placek, J.: Improvement of analytical accuracy of EPR spectrometer by taking into account variations in the shapes of samples. Fresenius' J. Anal. Chem. **343**, 863–872 (1992)

Nagy, V.Y., Sokolov, D.P.: Choosing reference samples for EPR concentration measurements. Part 2. Systems of S = 1. Anal. Chim. Acta **339**, 13–29 (1997)

Nagy, V.Y., Komozin, P.N., Desrosiers, M.F.: Choosing reference samples for EPR concentration measurements. Part 3. Systems of S = 3/2. Anal. Chim. Acta **339**, 31–51 (1997a)

Nagy, V.Y., Sokolov, D.P., Desrosiers, M.F.: Choosing reference samples for EPR concentration measurements. Part 4. Systems of S = 5/2. Anal. Chim. Acta **339**, 53–62 (1997b)

Narni, G., Mason, H.S., Yamazaki, I.: Quinhydrone as a quantitative standard for electron spin resonance spectrometry of biological systems. Anal. Chem. **38**, 367–368 (1966)

Noble, G.A., Markham, J.J.: Analysis of paramagnetic resonance properties of pure and bleached F-centers in potassium chloride. J. Chem. Phys. **36**, 1340–1353 (1962)

Ohno, K.: ESR imaging and its applications. Appl. Spectrosc. Rev. **22**, 1–56 (1986)

Ohno, K.: ESR Imaging. Magn. Reson. Rev. **11**, 275–310 (1987)

Piekara-Sady, L., Kispert, L.D.: ENDOR spectroscopy. In: Poole, J.C.P., Farach, H.A. (eds.) Handbook of Electron Spin Resonance, vol. 1. American Institute of Physics, New York (1994). Ch. V

Pierce, B.S., Elgren, T.E., Hendrich, M.P.: Mechanistic implications for the formation of the diiron cluster in ribonucleotide reductase provided by quantitative EPR spectroscopy. J. Am. Chem. Soc. **125**, 8748–8759 (2003)

Pilbrow, J.R.: Transition Ion Electron Paramagnetic Resonance. Oxford, London (1990)

Plato, M., Lubitz, W., Mobius, K.: A solution ENDOR sensitivity study of various nuclei in organic radicals. J. Phys. Chem. **85**, 1202–1219 (1981)

Poole, C.P.: Electron Spin Resonance: A Comprehensive Treatise on Experimental Techniques. Interscience, New York (1967)

Poole, C.P.: Electron Spin Resonance: A Comprehensive Treatise on Experimental Techniques, 2nd edn, p. 443. Wiley, New York (1983a)

Poole, C.P.: Electron Spin Resonance: A Comprehensive Treatise on Experimental Techniques, 2nd edn. Wiley, New York (1983b)

Poole, C.P., Farach, H.: Relaxation in Magnetic Resonance. Academic, New York (1971)

Poole, J.C.P., Farach, H. (eds.): Handbook of Electron Spin Resonance. AIP Press, New York (1994)

Poole, J.C.P., Farach, H.A.: Electron spin resonance. In: CRC Handbook of Spectroscopy, vol. II, p. 215 (1974)

Presley, T., Kuppusamy, P., Zweier, J.L., Ilangovan, G.: Electron paramagnetic resonance oximetry as a quantitative method to measure cellular respiration: a consideration of oxygen diffusion interference. Biophys. J. **91**, 4623–4631 (2006)

Rages, K.A., Sawyer, R.E.: Properties of microwave cavities containing magnetic resonant samples. Rev. Sci. Instrum. **44**, 830–834 (1973)

Randolph, M.: Quantitative considerations in electron spin resonance studies of biological materials. In: Swartz, H.M., Bolton, J.R., Borg, D.C. (eds.) Biological Applications of Electron Spin Resonance, pp. 119–155. Wiley, New York (1972a)

Randolph, M.L.: Quantitative considerations in electron spin resonance studies of biological materials. In: Swartz, H.M., Bolton, J.R., Borg, D.C. (eds.) Biological Applications of Electron Spin Resonance. Wiley, New York (1972b). Chap. 3

Rataiczak, R.D., Jones, M.T.: Investigation of the cw [continuous wave] saturation technique for measurement of electron spin-lattice relaxation. Application to the benzene anion radical. J. Chem. Phys. **56**, 3898–3911 (1972)

Rinard, G.A., Eaton, S.S., Eaton, G.R., Poole Jr., C.P., Farach, H.A.: Sensitivity in ESR measurements. Handb. Electron Spin Reson. **2**, 1–23 (1999a)

Rinard, G.A., Quine, R.W., Harbridge, J.R., Song, R., Eaton, G.R., Eaton, S.S.: Frequency dependence of EPR signal-to-noise. J. Magn. Reson. **140**(1), 218–227 (1999b)

Rinard, G.A., Quine, R.W., Song, R., Eaton, G.R., Eaton, S.S.: Absolute EPR spin echo and noise intensities. J. Magn. Reson. **140**(1), 69–83 (1999c)

Rinard, G.A., Quine, R.W., Eaton, S.S., Eaton, G.R.: Frequency dependence of EPR signal intensity, 248 MHz to 1.4 GHz. J. Magn. Reson. **154**(1), 80–84 (2002a)

Rinard, G.A., Quine, R.W., Eaton, S.S., Eaton, G.R.: Frequency dependence of EPR signal intensity, 250 MHz to 9.1 GHz. J. Magn. Reson. **156**(1), 113–121 (2002b)

Rinard, G.A., Quine, R.W., Eaton, S.S., Eaton, G.R.: Frequency dependence of EPR sensitivity. Biol. Magn. Reson. **21**, 115–154 (2004)

Robinson, B.H., Mailer, C., Reese, A.W.: Linewidth analysis of spin labels in liquids. I. Theory and data analysis. J. Magn. Reson. **138**, 199–209 (1999a)

Robinson, B.H., Mailer, C., Reese, A.W.: Linewidth analysis of spin labels in liquids. II. Experimental. J. Magn. Reson. **138**, 210–219 (1999b)

Russell, A.M., Torchia, D.A.: Harmonic analysis in systems using phase sensitive detection. Rev. Sci. Instrum. **33**, 442–444 (1962)

Saraceno, A.J., Fanale, D.T., Coggeshall, N.D.: An electron paramagnetic resonance investigation of vanadium in petroleum oils. Anal. Chem. **33**, 500–505 (1961)

Schell-Sorokin, A.J., Mehran, F., Eaton, G.R., Eaton, S.S., Viehbeck, A., O'Toole, T.R., Brown, C.A.: Electron spin relaxation times of fullerene ion (C_{60}^-) in solution. Chem. Phys. Lett. **195**, 225–232 (1992)

Schreurs, J.W.H., Fraenkel, G.K.: Anomalous relaxation of hyperfine components in electron spin resonance. II. J. Chem. Phys. **34**, 756–758 (1961)

Schreurs, J.W.H., Blomgren, G.E., Fraenkel, G.K.: Anomalous relaxation of hyperfine components in electron spin resonance. J. Chem. Phys. **32**, 1861–1869 (1960)

Schultz, S., Gullikson, E.M.: Measurement of static magnetization using electron spin resonance. Rev. Sci. Instrum. **54**, 1383–1385 (1983)

Schweiger, A.: Electron Nuclear Double Resonance of Transition Metal Complexes with Organic Ligands. Springer, Berlin (1982)

Segal, B.G., Kaplan, M., Fraenkel, G.K.: Measurement of g values in the electron spin resonance spectra of free radicals. J. Chem. Phys. **43**, 4191–4200 (1965)

Sert, I., Sünnetçioğlu, M.M., Surgur, R., Bingöl, G.: Dynamic nuclear polarization studies of TANOL/water-glycerol solutions. Z. Naturforsch. **55a**, 682–686 (2000)

Setaka, M., Sancier, K.M., Kwan, T.: Electron spin resonance investigation of electrical conductivity parameters of zinc oxide during surface reactions. J. Catal. **16**, 44–52 (1970)

Shaltiel, D., Genossar, J., Grayevsky, A., Kalman, Z.H., Fisher, B., Kaplan, N.: ESR in new high temperature superconductors. Solid State Commun. **63**, 987–990 (1987)

Sidabras, J.W., Mett, R.R., Hyde, J.S.: Aqueous flat cells perpendicular to the electric field for use in electron paramagnetic resonance spectroscopy, II: design. J. Magn. Reson. **172**, 333–341 (2005)

Siebert, D., Dahlem, J., Nagy, V.: Importance of transition probability values for accurate EPR concentration measurements. Anal. Chem. **66**, 2640–2646 (1994)

Slangen, H.J.M.: Determination of the spin concentration by electron spin resonance. J. Phys. E **3**, 775–778 (1970)

Smirnov, A.I., Norby, S.W., Walczak, T., Liu, K.J., Swartz, H.M.: Physical and instrumental considerations in the use of lithium phthalocyanine for measurements of the concentration of oxygen. J. Magn. Reson. B **103**, 95–102 (1994)

Smith, G.W.: Modulation effects in magnetic resonance: widths and amplitudes for Lorentzian and Gaussian lines. J. Appl. Phys. **35**, 1217–1221 (1964)

Sojka, Z., M. Che, M.: EPR Study of the Interation of O^{-}Ions with CH_3OH within the Coordination Sphere of Mo Ions Grafted in Silica: A New Approach for the Study of the Mechanism of Catalytic Reactions. J. Phys. Chem. **99**, 5418–5430 (1995)

Stesmans, A., VanGorp, A.: Improved measurement of the g factor of conduction electrons in lithium particles embedded in lithium-doped lithium fluoride. Phys. Lett. A **139**, 95–98 (1989)

Stoner, J.W., Szymanski, D., Eaton, S.S., Quine, R.W., Rinard, G.A., Eaton, G.R.: Direct-detected rapid-scan EPR at 250 MHz. J. Magn. Reson. **170**, 127–135 (2004)

Subramanian, S., Krishna, M.C.: Time-domain radiofrequency EPR imaging. In: Eaton, S.S., Eaton, G.R., Berliner, L.J. (eds.) Biomedical EPR – Part A: Free Radicals, Metals, Medicine, and Physiology. Biological Magnetic Resonance, vol. 23, pp. 321–383. Kluwer Academic, New York (2005)

Subramanian, S., Mitchell, J.B., Krishna, M.C.: Time-domain radio frequency EPR imaging. In: Berliner, L.J. (ed.) In Vivo EPR (ESR). Biological Magnetic Resonance, vol. 18, pp. 153–197. Kluwer Academic/Plenum, New York (2003)

Sünnetçioğlu, M., Bingöl, G.: An investigation of tanol/benzene solutions in weak and strong fields by EPR. Phys. Chem. Liq. **26**, 47–58 (1993)

Sünnetçioğlu, M., Bingöl, G., Sungur, R.: Hyperfine structure and relaxation times of 4-Oxo-TEMPO/methyl alcohol solutions in weak and strong fields. Z. Naturforsch. **46a**, 976–982 (1991)

Swallen, S.F., Kearns, K.L., Mapes, M.K., Kim, Y.S., McMahon, R., Ediger, M.D., Wu, T., Yu, L., Satija, S.: Organic glasses with exceptional thermodynamic and kinetic stability. Science **315**, 353–356 (2006)

Swartz, H.M., Iwasaki, A., Walczak, T., Demidenko, E., Salikhov, I., Khan, N., Lesniewski, P., Thomas, J., Romanyukha, A., Schauer P, D.S.: In vivo EPR dosimetry to quantify exposures to clinically significant doses of ionizing radiation. Radiat. Prot. Dosimetry **120**, 163–170 (2006)

Tipikin, D.S., Lazarev, G.G., Lebedev, Y.S.: Mechanochemical generation of stable radical pairs. Z. Fizich. Khim. **67**, 176–179 (1993)

Tipikin, D.S., Levbedev, Y.S., Rieker, A.: Mechanochemical generation of stable radical species. Oxidation of pyrocatechols. Chem. Phys. Lett. **272**, 399–404 (1997)

Towell, J., Kalyanaraman, B.: Detection of radical adducts of 5, 5-dimethyl-1-pyrroline N-oxide by the combined use of high-performance liquid chromatography with electrochemical detection and electron spin resonance. Anal. Biochem. **196**, 111–119 (1991)

Tseitlin, M., Dhami, A., Quine, R.W., Rinard, G.A., Eaton, S.S., Eaton, G.R.: Electron spin T_2 of a nitroxyl radical at 250 MHz measured by rapid scan EPR. Appl. Magn. Reson. **30**, 651–656 (2006)

Tseitlin, M., Dhami, A., Eaton, S.S., Eaton, G.R.: Comparison of maximum entropy and filtered back-projection methods to reconstruct rapid-scan EPR images. J. Magn. Reson. **184**, 157–168 (2007)

Urbanski, T.: Formation of solid free radicals by mechanical action. Nature **216**, 577–578 (1967)

Vigouroux, B., Gourdon, J.C., Lopez, P., Pescia, J.: Broadening of the ESR line, caused by the variation of the cavity Q factor across the magnetic resonance. J. Phys. E. **6**, 557–558 (1973)

Wada, T., Yamanaka, M., Fujihara, T., Miyazato, Y., Tanaka, K.: Experimental and theoretical evaluation of the charge distribution over the ruthenium and dioxolene framework of [Ru (OAc)(dioxolene)(terpy)] (terpy) 2, 2':6', 2''-terpyridine) depending on the substituents. Inorg. Chem. **45**, 8887–8894 (2006)

Wahlquist, H.: Modulation broadening of unsaturated Lorentzian lines. J. Chem. Phys. **35**, 1708–1710 (1961)

Warren, D.C., Fitzgerald, J.M.: Parameters influencing the electron spin resonance signal intensity of metal ion complex-exchanged resins. Anal. Chem. **49**, 1840–1842 (1977)

Watanabe, T., Sasaki, T., Fujiwara, S.: Phase dependence of saturation transfer EPR signals and estimated rotational correlation times. Appl. Spectrosc. **36**, 174 (1982)

Weger, M.: Passage effects in paramagnetic resonance experiments. Bell Syst. Tech. J. **39**, 1013–1112 (1960)

Weil, J.A., Anderson, J.K. The determination and reaction of 2,2-diphenyl-1-picrylhdrazyl with thiosali-cylic acid. J. Chem. Soc. 5567–5570 (1965)

Weil, J.A., Bolton, J.R.: Electron Paramagnetic Resonance: Elementary Theory and Practical Applications, 2nd edn. Wiley, Hoboken, NJ (2007)

Weil, J.A., Bolton, J.R., Wertz, J.A.: Electron Paramagnetic Resonance: Elementary Theory and Practical Applications, p. 497. Wiley, New York (1994a)

Weil, J.A., Bolton, J.R., Wertz, J.E.: Electron Paramagnetic Resonance: Elementary Theory and Practical Applications. Wiley, New York (1994b)

Wertz, J.E., Bolton, J.R.: Electron Spin Resonance: Elementary Theory and Practical Applications. McGraw-Hill, New York (1972)

Westenberg, A.A.: Use of ESR for the quantitative determination of gas phase atom and radical concentrations. Prog. React. Kinet **7**, 23–84 (1975)

Wilson, G.V.H.: Modulation broadening of nuclear magnetic resonance and electron spin resonance line shapes. J. Appl. Phys. **34**, 3276–3285 (1963)

Wright, A.C., Song, H.K., Wehrli, F.W.: In vivo MR micro imaging with conventional radiofrequency coils cooled to 77° K. Magn. Reson. Med. **43**, 163–169 (2000)

Yordanov, N.D.: Quantitative EPR spectroscopy – "state of the art". Appl. Magn. Reson. **6**, 241–257 (1994)

Yordanov, N.D.: Is our knowledge about the chemical and physical properties of DPPH enough to consider it as a primary standard for quantitative EPR spectrometry. Appl. Magn. Reson. **10**, 339–350 (1996)

Yordanov, N.D., Christova, A.G.: DPPH as a primary standard for quantitative EPR spectrometry. Appl. Magn. Reson. **6**, 341–345 (1994)

Yordanov, N.D., Christova, A.G.: Quantitative spectrophotometric and EPR-determination of 1, 1-diphenyl-picryl-hydrazyl (DPPH). Fresenius' J. Anal. Chem. **358**, 610–613 (1997)

Yordanov, N.D., Genova, B.: Analysis of some non-linear effects in quantitative electron paramagnetic resonance spectrometry. Non-linear effects due to the use of cavities with TE_{102} and TM_{110} modes. Anal. Chim. Acta **353**, 99–103 (1997)

Yordanov, N.D., Ivanova, M.: The present state of quantitative EPR spectrometry: the results from an international experiment. Appl. Magn. Reson. **6**, 333–340 (1994a)

Yordanov, N.D., Ivanova, M.: Test studies of some conventional CW EPR spectrometer settings. Appl. Magn. Reson. **6**, 347–357 (1994b)

Yordanov, N.D., Slavov, P.: Influence of the diameter and wall thickness of a Quartz pipe inserted in the EPR cavity on the signal intensity. Appl. Magn. Reson. **10**, 351–356 (1996)

Yordanov, N.D., Gancheva, V., Pelova, V.A.: Studies on some materials suitable for use as internal standards in high energy EPR dosimetry. J. Radioanal. Nucl. Chem. **240**, 619–622 (1999)

Yordanov, N.D., Mladenova, B., Petkov, P.: Studies on the uncertainties in quantitative EPR estimations due to the construction of the cavities used. Anal. Chim. Acta **453**, 155–162 (2002)

Zecevic, A., Eaton, G.R., Eaton, S.S., Lindgren, M.: Dephasing of electron spin echoes for nitroxide radicals in glassy solvents by non-methyl and methyl protons. Mol. Phys. **95**(6), 1255–1263 (1998)

Zweier, J.L., Samouilov, A., Kuppusamy, P.: Cardiac applications of in vivo EPR spectroscopy and imaging. In: Berliner, L.J. (ed.) In Vivo EPR (ESR), vol. 18, pp. 442–468. Kluwer Academic/Plenum, New York (2003)

Index

GPSR Compliance

The European Union's (EU) General Product Safety Regulation (GPSR) is a set of rules that requires consumer products to be safe and our obligations to ensure this.

If you have any concerns about our products, you can contact us on ProductSafety@springernature.com

In case Publisher is established outside the EU, the EU authorized representative is:

Springer Nature Customer Service Center GmbH
Europaplatz 3
69115 Heidelberg, Germany

Batch number: 10726493

Printed by Printforce, the Netherlands